No. 2061
$12.95

COLLECTING AND RESTORING OLD STEAM ENGINES

BY RICHARD J. EVANS

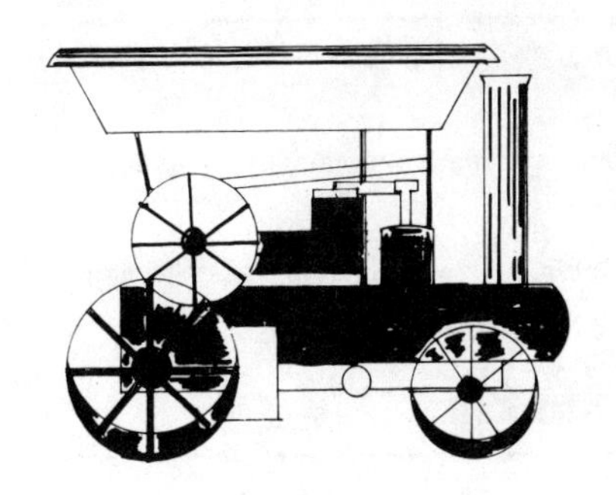

MODERN AUTOMOTIVE SERIES
TAB BOOKS Inc.
BLUE RIDGE SUMMIT, PA. 17214

FIRST EDITION

FIRST PRINTING—MAY 1980

Printed in the United States of America

Library of Congress Cataloging in Publication Data

Evans, Richard J. 1935-
Collecting and restoring old steam engines.

Includes index.
1. Steam-engines—Collectors and collecting.
I. Title.
TJ469.E94 621.1'028 79-23501
ISBN 0-8306-9718-7
ISBN 0-8306-2061-3 pbk.

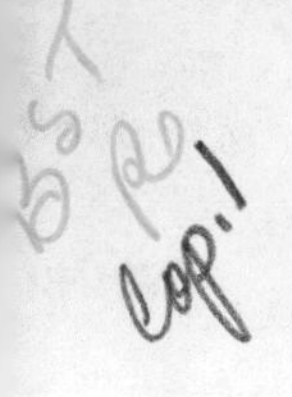

Acknowledgments

The author wishes to acknowledge the immense contribution of Graham Warhurst, sometime Chief Engineer of the Seaton Tramway system and the Isle of Man Steam Railway, now Assistant Operations Manager of the Isle of Man Steam & Electric Joint Railways Board. Graham has not only corrected the author's efforts but also added considerable wordage of his own accord, without which the book would not only be less complete than it now is, but could not have been written at all.

We gratefully acknowledge the help of

The Science Museum of London, England

which has provided the majority of the photographs used in this book. We acknowledge additional material from

The Museum of Transportation,
Brookline, Boston, MA.

The Manx Motor Museum,
Crosby, Isle of Man, British Isles
and
Peter J. English, Esq. & Graham Warhurat, Esq.

The author also wishes to thank: British Steam Specialities of Leicester; Pegler-Hattersley of Doncaster; Stuart-Twiner of Henley-on-Thames; and G. & A. E. Slingsby of Hull for some of the photographs and illustrations depicting equipment used in steam plants.

Contents

Introduction

The romance of steam is inexorably tied up with the railroad. There can be few men over the age of 40 who do not remember the large hissing, clanking, steaming monsters, bells shrilling as they trundled and rolled their way along the long steel road.

However, railroads are only one of the many uses of steam power. Equally important, but without the glamour, industry largely owes its success to the steam engine. Before the steam engine the only easily available source of rotative power was the water wheel.

Once steam engines became available they were adapted to a wide range of commercial and domestic uses, so that today the would-be restorer has a great variety of possible sources from which to find one.

This book ignores entirely railroad engines, but discusses in depth, or in passing, many of the other uses for the steam engine. Its prime purpose is to offer a guide to restoration, not only of the engine itself, but also of the major and minor ancillaries that are an essential part of the power source.

It is not the aim to provide a comprehensive textbook for every aspect, but merely to give a broad outline of the problems, to enlarge on those items which it is thought may present the most difficulties, and, hopefully, to stimulate the reader to search further in this field of industrial science.

Those who seek the most information will have the greatest pleasure and meet the least problems in their restoration work.

The author and his collaborator are both residents of the Isle of Man, a Victorian enclave, where if wives interfere with hobbies, wives, not the hobbies, are changed! We hope to educate our readers along similar lines!

STEAM CAN BE DANGEROUS

It is assumed that in order to be sufficiently interested to purchase this book, the reader has a basic knowledge of steam engines and the dangers involved with a steam plant. Nevertheless, it cannot be stressed too strongly that STEAM CAN BE DANGEROUS and that any advice given in this book must be used in conjunction with the particular conditions appertaining to the restoration or operation being carried out, and amended for these conditions if necessary. No responsibility by author or publisher is accepted for acting on this advice.

By Richard J. Evans

Chapter 1
In The Beginning

A popular myth is to the effect that James Watt invented the steam engine while watching a kettle boil at his home and observing the force with which a boiling kettle ejected a cork jammed in the spout with the lid tied down.

In reality, the steam engine was developed over a relatively long period and was only made possible by the advances made in various forms of science, including metallurgy. Iron and steel had to have reached the point in development where boilers could be made that would not explode and cylinders cast and bored sufficiently accurately so that pistons could make a reasonably satisfactory seal therein.

The basic principle of a steam engine is that water when boiled gives off a vapor which tries to expand and, in so doing, applys pressure. This steam is channeled into one end of a steel cylinder and pushes on a piston forming the other end, such piston being free to slide out of the cylinder, and connected by mechanical links to some form of mechanism. To make the piston sufficiently tight in the cylinder that the steam will not escape down its sides, yet not being so tight it will not move, piston rings are used.

Thus, this power source is basically a simple mechanism. The complexity comes in the accessories required to make the correct amount of steam for the work to be done and to lubricate the moving parts. Production of steam depends upon supplying water and heat in the correct quantities and at the right time. Such heat can be generated either by solid or liquid fuel or by electricity.

The steam engine took about 100 years to reach its zenith, whereas the internal combustion engine attained a very reliable state in less than 50 years from its inception.

It is generally accepted that the Frenchman Cugnot built the first steam powered road vehicle around 1770—a gaint wood-framed, three-wheel monster with a circular front-mounted boiler overhanging the single front wheel. This vehicle can still be seen in the Conservatoire des Arts and Metiers in Paris, having probably only been used twice in its life. The second time it overturned. (See Fig. 1-1). This was followed by the wheeled boat built by Evans in America, circa 1800, and was only made to travel on the road as a means of getting it to its launching place. After this, railroad engines started to appear, together with increasing use of steam power in factories and agriculture for driving every conceivable form of machinery.

By 1930, the steam road vehicle was nearly dead and steam in factories was dying. Thus, the steam era can be considered as a 150 years, 1800 to 1950, after which even steam railroad engines were becoming extinct in the Western world and declining elsewhere.

It is unlikely steam power will ever again ascend over the internal combustion engine unless nuclear fuels become practical, as it is an inherently less efficient way of extracting power from fossil fuels.

Earliest Practical Passenger Carrier

Probably the earliest practical passenger-carrying vehicles were Trevithick's 1801/1803 designs. His 1802 carriage (Fig. 1-2) has been well documented and is often mistakenly thought to have only three wheels. In fact it has two close together at the front, with center-axle steering pivot and a tiller—which can not have helped ease of steering! Its single cylinder lay horizontally fore and aft, behind the rear axle, with a long piston rod coming forward over this axle to a crank on a countershaft running in plumbing blocks amidships. Spur gears on the ends of this shaft mesh with larger spur gears fixed to the rear wheels. The countershaft and corresponding wheel gears are of differing sizes each side of the car, either side being able to be clutched to the countershaft independent of the other, thus giving a choice of gear ratio. Neither front nor rear springing appears to be provided. These vehicles operated in London for some time until the designer transferred his interest to railroad traction.

In 1825, Thomas Blanchard built what is reputedly America's second steam road vehicle—virtually a wagon—but this was really only a toy. The most serious steam vehicle operator of this period was Goldsworthy-Gurney whose fleet of coaches worked around London and, occasionally, as far as Bath—about 150 miles to the west. (See Fig. 1-3).

Another eminent London operator a few years later was Walter Hancock, who ran a series of quite well-designed steam omnibus with boilers, not unlike central heating radiators in appearance. The style of these "buses" was quite modern for their date; they could well have passed for those of 50 years later, having enclosed passenger accommodation with glass windows, the engine slung below the body, the boiler at the rear and having the "conductor," or driver, seated at the very front, it

Fig. 1-1. The Cugnot steam tractor of 1770 is surely as well-known to everyone as is the tale that it crashed on its second and ultimate outing, being driven into a wall. This particular view may, however, be new to some, with a scale in inches to show dimensions.

was one of the earliest examples of forward control. Steering was by means of a wheel and chain to the center-pivoted front axle. (See Fig. 1-4.) The main problems with these early public transport ventures were two. First, metallurgy had not progressed to the point where reliable axles, bearings and other components could be produced; and, secondly, there was the age-old reaction of so much of the public to anything new—it must be the work of the devil.

In 1866, a Ware steamer was exported from Bayonne, N.J., to Prince Edward Isle, then a U.K. Crown Colony. According to the *Scientific American* of 1867:

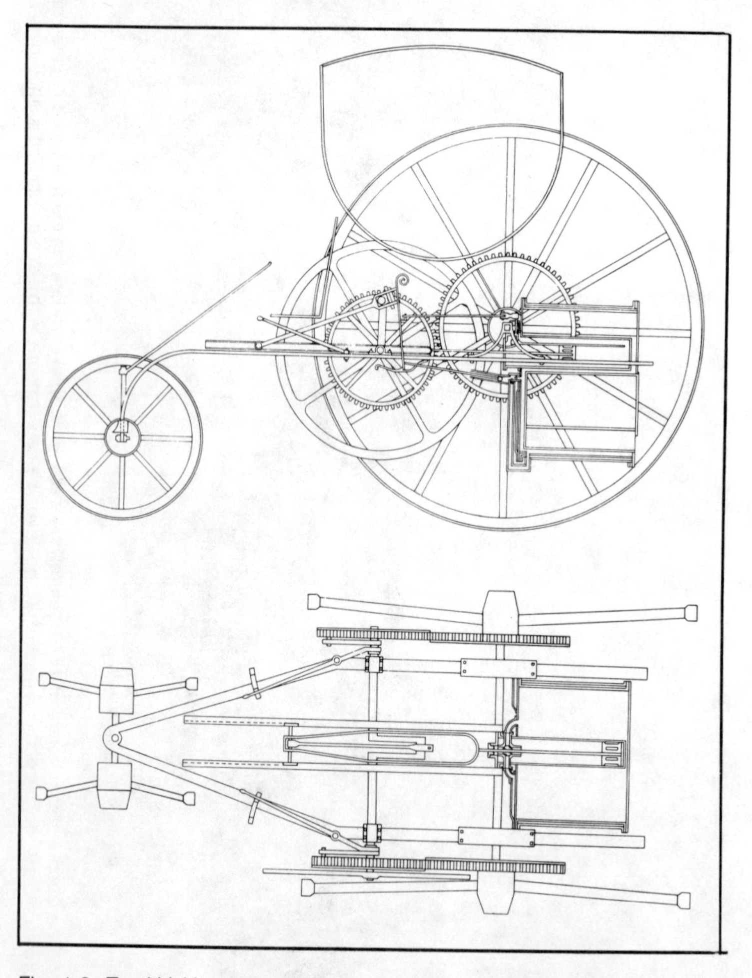

Fig. 1-2. Trevithick's very early steam carriage of 1802 really worked, and reasonably well, but the designer lost interest in road vehicles. It had a horizontal engine with a single cylinder pointing to the rear.

Fig. 1-3. Church's highly elaborate early steam road vehicle had exquisite ornamentation, but was highly impractical. Did it, in fact, ever exist?

"It presents a graceful appearance—the boiler is hung between the forks of a frame of steel, which meet on the forward axle and thence backward diverge, holding the boiler suspended in the triangle thus formed. This frame of steel, edge up, is twisted a half turn on each side of the boiler, thus acting as a spring. The engines (plural?) work on an incline and drive a shaft with a chain wheel, which, by a machine chain, rotates the driving shaft and wheels. The engine is intended to give three revolutions to the first shaft to one revolution of the driving wheels, thus gaining power for ascending inclines. The differences can be multiplied to nine times. A lever in front of the drivers seat serves, by a simple mechanism, to guide the machine when used as a carriage, and a rod and handle connected to the engine shaft readily reverses the motion of the engine."

An interesting extract, this quotation shows the type of information disseminated to the public—rather more detail than most sales catalogues offer today. The only problem is that, even with modern scientific knowledge, the author does not understand very much of the description; thus, one is skeptical about how much the average reader, with scientific leanings 100 years ago, could understand.

Early Steam vs. Early I.C. for Cars

The reason for the popularity of the steam car at the turn of the century is not hard to find. The internal combustion engine was in its infancy and, lacking development, sophisticated metals or a reliable ignition system, was not a very satisfactory power unit. It tended to run with moderate reliability at only one speed and required a gearbox to translate that speed to that required for power purposes—not too much of a problem for a stationary plant, but a matter of considerable difficulty when the varying speeds of road vehicles had to be met. Whereas, in the words of a contemporary journalist, 'The steam car keeps chugging along, in town, out of town, uphill, downhill, at any speed of the conductors whim, with nowt but a slight movement of one lever." In reality, a gross exaggeration; but, the basic idea of no gears to change and a variable speed engine was correct.

However, these reasonably sophisticated steam cars did not appear overnight from the earlier experiments. Steam enthusiasts had a hard time during the mid-period of the nineteenth century. Public reaction to breakdowns, swinging charges by turnpike operators, and objections from the horsing fraternity resulted in most of the early experimenters giving up public use of steam buses, and development dissolved into on-off private ventures such as the Tricar of Catley and Ayres in 1868. (See Fig. 1-5.)

This funny little vehicle had a single front wheel, and was tiller-steered, with two driving wheels at the rear; but, in this case, all wheels sprung—the front by an early example of coil springing and the rear by the conventional half-elliptic "carriage" spring. The boiler was of the vertical type situated at the rear, apparently with the rear axle running through the firebox.

Apart from these on-off experiments, the steam road scene then towards the end of the 1800's moved to France with Bollée family to the fore, followed by Comte de Dion, who produced a variety of highly practical road vehicles, the larger of which carried more than a dozen people—in this case, by means of a steam tractor unit towing a trailer "omnibus."

Fig. 1-4. One of Hancock's well-known Omnibus, circa 1833—A rather romantic engraving—but these buses did have some measure of success in London during the 1830's, and if public opinion had not, as usual, condemned anything new, could well have developed into a national bus service very much earlier than, in fact, occurred.

Fig. 1-5. The Catley & Ayres 1868 steam tricycle was one of the better earlier attempts, and had metallurgy and the highways been better, could well have enjoyed considerable popularity.

Many manufacturers made steam stationary plants and, in due course, branched out to produce traction engines. These machines developed from wheeled stationary plants or "portable engines," as they were generally called. The transition from portable to self-moving was achieved initially by chain drive from the engine crankshaft to the wheels, any steering being carried out by a horse in shafts at the front who frequently had to pull the thing, as well, when the chain broke or fell off. Later products used an all-gear drive and abandoned horse steering in favor of a chain and worm system with a wheel adjacent to the engineer that, upon turning, eventually and laboriously turned the center-pivoted front axle.

The natural progression from traction engines was to wagons, or "lurrys," as they were usually called. Many of the later well-known truck makers started life with steam. For example, White, Foden, Locomobile, Leyland—most of these had forsaken steam within a few years of the end of the Great War.

Steam vehicles for private use were mainly of the light buggy type epitomized by the Stanley/Locomobile design which, with slight variations to avoid patent infringements, was widely copied by literally dozens of firms. Without doubt, the U.S.A. was the stronghold of steam cars, although most makes had succumbed by the end of the First World War. Stanley kept going until the mid-Twenties. Doble actually started his "E" series in the Twenties, running them through into the Thirties; but, although these, with the "F", were the best steam cars ever made, only a very small number were produced—reputedly 42 in all.

In Europe the extent of the steam car after 1900 was mainly Serpollet in France and Turner-Miesse in the U.K., although this latter owed much to its Belgian parentage.

Steam was more successful for commercial use than for private purposes. This was chiefly due to its inherent design characteristics, namely that it lent itself best to slow heat up, long heat retention principles. Flash boilers worked after a fashion, but had little reserve of steam; whereas, if one had plenty of room, as in a wagon, for a large boiler and large fire, a good head of steam could be built up to cope with any extra demands such as acceleration or hills. The snag was a long warm up time. To fire up a cold firetube boiler steamer to full working pressure could take an hour, or considerably more for a big unit, but it could then be kept hot all day with little effort—ideal for a vehicle in continuous use conveying loads, performing stop and start tasks, or for traveling long distances, but not so good for a car on shopping trips.

Steam Handicaps Obvious Early

There was also a psychological problem in that the upper classes, who could afford steamers, had a marked reluctance to enter Heaven or, more likely, Hell at an early date due to violent misbehavior of the roaring fire and high pressure steam container on which they were sitting.

At this point we will reiterate the warning which appears at the beginning, of this book and which we will repeat again at intervals—**"Steam is Dangerous"**. If something goes wrong with an internal combustion engine, the worst that is likely to happen is a wrecked motor with, possibly, the odd bit flying through the engine cover. If you are careless with steam or something gives way, you may be the person flying—and probably not all in one piece. Seriously, it cannot be emphasized too strongly that steam can kill.

There were also operational problems in that, although the majority of a steam vehicle's mechanics were reliable, liquid-fuel burners are notoriously troublesome. If condensers were not used, 20 to 30 miles was the usual range between water tank refills. If you had a condensing vehicle, your problem often was cylinder oil getting carried back to the boiler and insulating the crown plate, thereby causing it eventually to collapse with the inevitable explosion.

Appeal of Stationary Steam Engines

The majority of stationary engines were single-cylinder designs with large flywheels and a belt drive to various types of machinery. (See Figs. 1-6 and 1-7.) Although of the same basic principles, the size of engines ranged from about 5 hp driving, say, a mixer at a small rural brickworks, to

Fig. 1-6. Boulton & Watt's rotative beam ("Lap") engine of 1788 was a very early stationary engine of the sort used in the early stages of the English Industrial Revolution. Note the gear method of transmitting the drive, rather than the crank.

Fig. 1-7. A single-cylinder horizontal steam engine made by the English firm of Ruston-Hornsby, the makers of one of the better traction engines—note the large flywheel required with a single-cylinder engine.

a two-cylinder machine producing 2000 hp and powering an entire cotton spinning and weaving mill.

The size of engine likely to be encountered by the "preservationist restorer" is from the smallest up to about 100 to 150 hp. (See Fig. 1-8.) This type of engine, with most of the working parts in the open, as it were, makes a very satisfactory sight slowly turning over on just a breath of steam. (See Fig. 1-9.) Although, as mentioned earlier, the steam engine was commercially outmoded by the 1930's, many industries which had steam supplies readily available, or in which steam was part of a necessary process, retained and even renewed steam engines as the most convenient, if not cheapest, prime movers much later than this.

Examples of such late use were to be found until quite recently in such places as laundries, chemical plants, dye works and small saw mills. In the case of the latter, the attraction was a readily available "free" fuel supply in the form of sawdust and wood chips. Large numbers of small efficient boilers were also produced for various purposes in the Thirties and Forties, probably the best known being the "Merryweather." The acquisition of a boiler of this type will complete your own private steam plant!

As can be seen in the picture of the Robey Portable Power Plant (Fig. 1-10), these early wheeled engines were very similar to the subsequent traction engines, it being but a short step to add a platform for the driver and some form of crude drive and steering, as mentioned earlier in this chapter.

The English developed traction engines far faster than the Americans. Compare the English 1871 Aveling & Porter 8 hp engine with its horizontal boiler and gear-driven rear wheels with wide treads for soft ground with the American 1877 Oshkosh that completed in the Wisconsin 1878 race for "Steam Wagons." This has a vertical boiler making viewing of the road ahead most difficult, narrow wheels that tend to bog down in

Fig. 1-8. Small single-cylinder vertical steam engine with feed pump—a typical engine used for small power requirements.

Fig. 1-9. An example of the use to which a small steam engine can be put—a Willans engine of 1885 driving a dynamo. This particular engine has central valve gear, as can be seen. A well restored piece of equipment such as this makes an extremely attractive exhibit.

soft going, a very crude chain drive and primitive accommodation for the driver. (See Fig. 1-11.)

As the years progressed, the traction engines slowly improved with, in many cases, the addition of canopies affording some protection from the elements and, depending on the main use to which a particular example was put, solid rubber tires or, very occasionally, pneumatic tires. (See Fig. 1-12.)

A comparatively rare type, but that which is most sought after today, was the Showmans Engine. This was an elaborately decorated, canopied engine, usually with a belt driven generator mounted on a platform above the smoke box and ahead of the smoke stack.

Steam Rollers, Steam Wagons and Steam Motorcycles

Once tar macadam roads started to proliferate, the traction engine soon spawned the steam roller, in which the leading pair of wheels of the traction engine were replaced by a single, or duplex, roller. (See Figs. 1-13 and 1-14.) Some of these machines had disc rear wheels, also wide and flat for rolling compared to the more normal spoked wheel of their forebears. A word of warning here: There are one or two so-called traction engines which are clever conversions of road rollers and not the genuine article at all.

The next progression was the steam wagon or lorry. In some cases, as for example the Foden, (See Fig. 1-15) these were similar to the traction engine with the rear wheels moved back some distance and an open truck type of body interposed behind the driver's platform. These

Fig. 1-10. Robey's portable steam engine—the similarity to a locomotive boiler is obvious, it being but a small step to provide a wheel to steer the front axle and a chain to drive the rear wheel.

Fig. 1-11. An Aveling & Porter traction engine of 1871 was, by that time, the familiar pattern of the traction engine, was well established and there were few major variations until its death as a commercial proposition.

vehicles, of course, had higher gearing and either solid or pneumatic tires. Others were purposely designed as lorries, such as the Sentinel which started before the Great War and continued into the late Thirties. These always had a vertical boiler but, fortunately, the top of this was well below the driver's eye-line. Thus, the only thing that impeded forward vision was a relatively small-diameter chimney or smoke stack. This boiler position had great advantages in winter, but equally great disadvantages in summer. For the majority of the life of this make, the boiler working pressure remained at 230 p.s.i., steam being raised by coal or coke to feed a twin-cylinder horizontal engine which, in turn, drove the rear wheels by chain. In later years, all wheels were shod with pneumatic tires.

Quite a large number of firms made steam wagons and so reliable were they that many firms continued using them into the Fifties, especially where coal was easily or cheaply obtainable and the operator's trade demanded a vehicle in more or less continuous use throughout the working day. In the ten years from 1913 to 1922, including a four-year war period in Europe, 17 different makes were made in England alone. None of these, incidentally, had pneumatic tires. (See Fig. 1-16.)

At the other end of the steam road vehicle spectrum is the motorcycle saga. Steam motorcycles were tried as early as three or four-wheeled vehicles. For example, the 1869 effort of Michaux Perrioun, which is virtually a Hobby Horse "Bicyclette," had a weird and wonderful

Fig. 1-12. A Ruston-Hornsby of 1920 was one of the last of the traction engine species, the only obvious major difference from the 1871 Aveling & Porter being solid rubber tires for highway use.

Fig. 1-13. A selection of pre-World War I road rollers of various makes in use for road-mending in India during 1977.

steam device inserted between the frame and saddle. If this worked at all, which is doubtful, it drove the rear wheel by means of a long belt, but the provision of pedals on the front wheel would seem to offer a more reliable and safer means of propulsion. (See Fig. 1-17.) Would you like to have a

Fig. 1-14. Marshall Steam Road Roller still in daily use by the National Asphalt & Construction Co. of Chembur, India, proves that even in a dilapidated state, steam vehicles will usually run.

Fig. 1-15. Foden Steam Lorry (1910) with a locomotive type boiler left less load space than the vertical boiler of its contemporary, the Sentinel.

Fig. 1-16. Sentinel steam wagon of 1920 was rated at six tons load with underfloor engine and vertical coal-fired boiler. This was one of the last steam lorries to be made in England, and many were in use, particularly by breweries, into the late Fifties.

roaring fire within inches of a rather personal part of your anatomy? Also about this time in America, one Silvester Roper mounted a steam engine on a velocipede. This engine had a cylinder on either side of the cycle's backbone from which long connecting rods drove cranks directly on the

Fig. 1-17. The Michaux Perrioux steam motor cycle of 1869 was one of the weird and wonderful early European experiments. The pedals on the front wheel were undoubtedly essential!

Fig. 1-18. The Hildebrand Steam Motor Cycle of 1889—some progress had been made by this date, the basis of the machine being very obviously the Safety Pedal Bicycle. Range was limited to the amount of heat that could be obtained from fuel in the burner with no provision for recharging with fuel.

rear wheel axle. The coal-fired boiler hung between the wheels with a raked funnel behind the saddle—ideal for keeping the rear end of the engineer warm!

Twenty years later Hilderbrand produced a somewhat safer device, a more or less conventional bicycle frame in which the space between top and bottom bars, between one's legs, was filled with a water tank; the boiler and burner being not only below this, but also below one's feet. A single-cylinder engine appeared to drive direct to the rear wheel with a valve-operating eccentric drive on an extension of the wheel spindle or crankshaft outside the frame. It would appear that the range of this machine was somewhat limited, as the boiler was fired by coal which could not be replenished while on the move. (See Fig. 1-18.)

Chapter 2 Twentieth Century Private Vehicles

By the first decade of the twentieth century, several further attempts at steam motorcycles had been undertaken, one of the last being the Pearson-Cox which was on the market late in 1912. (See Fig. 2-1). This machine had a flash boiler consisting of a 65-foot coil of 1/4-inch bore steel tube heated by a liquid fuel burner, the tank for which was mounted over the rear wheel—in the fashion of the later weird internal combustion engined device, the five-wheeled Briggs & Stratton. The water was contained in a triangular tank in the frame between this rider's legs. The engine was located vertically between this water tank and the rear wheel, having a stroke of two inches and a bore of one and three quarters inches. This was of the single-acting type, with a chain drive to the rear wheel giving a ratio of three and one half turns of the engine to one of the rear wheel. Front springing was by twin coils, the rear wheel being unsprung.

Relatively few two-wheeled steam vehicles were manufactured. A few were produced with three wheels, but the overwhelming majority had four wheels.

Serpollet and Gardener

Among the three-wheelers was Serpollet's 1889 steam carriage which, in a way, was a rather improved version of Cugnot's vehicle of over 100 years earlier. (See Fig. 2-2.) It was the same layout of artillery wheels, one in front, two at the back, but with chain drive from the midships engine back to the rear wheels. Steering was by a single-spoked handlebar, with dual full-elliptic springing on the front wheel, whereas the rear was sprung on conventional semi-elliptic "cart" springs. The car was reasonably well-balanced with the vertical boiler right at the rear. This

Fig. 2-1. Pearson-Cox Steam Motor Cycle of 1913, showing the boiler in section. This is one of the water tube "Flash" types. The tank over the rear mudguard holds the fuel; the tank between the driver's legs is for water.

was of the flash type in which only the amount of steam actually needed at a particular time was generated, thus enabling steam to be raised far more quickly. The marque continued in production, changing its name in the process to Gardener-Serpollet, until 1907 and contributed a great deal to the development of the passenger steam vehicle.

Fig. 2-2. The Serpollet Steamer of 1889 had a rear vertical boiler, cart type wheels and single-wheel front steering.

Within a few years Serpollet abandoned coke as a fuel for the far more easily controlled paraffin (kerosene). It was comparatively easy to use a pair of pumps to feed fuel and water to the boiler in suitable proportions for the varying steam requirements of the engine. By 1899 he had a 5 hp engine with four vertical cylinders and poppet valves instead of slide valves. In 1900, the American, Mr. Gardener, supplied the finances and joined his name to the company to enable production to be substantially increased, making this one of the most numerous of European steam cars. (See Fig. 2-3.)

Another leading European steam vehicle maker was De Dion who, with Georges Bouton, produced various steam road vehicles starting with a tricycle late in the nineteenth century and continuing until 1904. (See Fig. 2-4.) Most of the vehicles were produced during the last ten years of this period, being more akin to buses than private cars, as their far better known internal combustion engine was now the leading motive power for the latter.

Stanley Leads Steam Popularity

Once the new century had gotten into its stride, steam cars began to proliferate, particularly in America. The Stanley twins had sold out their design within a short period of its birth in 1897 and subsequently bought all the patents back for a tenth of the selling price, less than five years later! This marque was the backbone of all steam car production, running as either Stanley or Locomobile, for 30 years. (See Fig. 2-5.) All this time the Stanleys used a firetube boiler (See Fig. 2-6.) and did not add a condenser until 1915, with the addition of which, the familiar "coffin-nose" front gave way to an appearance similar to that of most other American cars, gasoline or steam, of the era. (See Figs. 2-7 and 2-8.)

Among the numerous makers that started out with steamers was the Foster, manufactured in Rochester, N.Y., which was a cross between a Locomobile and a CX Stanley in design. (See Fig. 2-9.)

Another of the better known American makes was the White. This started out in 1900 with a twin-cylinder engine with a condenser fitted in a surrey type body. By 1903, the famous front-mounted compound engine had been adopted, initially with Stephenson link valve motion and from 1909 onwards replaced by Joy valve gear. (See Fig. 2-10.) From the onset of the compound engine, a flash boiler was used and the marque's customers included no less a person than Theodore Roosevelt. By 1912, the company had forsaken steam for gasoline cars, giving up all car production, except for special order, by 1919 to concentrate on trucks.

The main English contender in the steam car field was not really English at all, but a Belgian design made under license, as mentioned in the last chapter. This was the Turner-Miesse, produced from 1902 to 1913 and consisting of a three-cylinder, single-acting engine with chain drive and a front-mounted, flash-type boiler with condenser. Horse powers ranged

Fig. 2-3. A Gardner-Serpollet, circa 1903, with detailing of water tube "Flash" boiler and top view of the engine—it will be seen that this is an horizontally opposed unit, mounted low down in the chassis, with valves operated by a camshaft very much akin to modern gasoline engine practice. This French-American design was one of the better examples of the early steam car. Scale on the water tubes can be seen. This could either build up from products of combustion, or be flaking of the tubes due to heat.

Fig. 2-4. The De Dion tricycle, made in the late nineteenth century and now on view at the Museum of Transportation, Boston, MA., was made in France by Comte de Dion, one of the early successful steam vehicle manufacturers.

Fig. 2-5. This Locomobile, circa 1900, is the Stanley early design purchased by the Locomobile Company and made by them for a few years. This design was widely copied by other makers in its basic principles. The engine is vertical behind the flap below the front seat, with the boiler behind.

Fig. 2-6. This Stanley Steamer of 1904 type CX of 10 hp with 16-inch diameter copper tube boiler was restored in 1953 and was recently on show in the Meridith Auto Museum at Meridith, NH. It is reputed to cover 0 to 45 mph in ten seconds, if the gods are favorable!

Fig. 2-7. This Stanley Gentlemans Speedy Roadster is at the Hershey A.A.C.A. annual meet, being judged in 1978. This is reputed to be the most desirable Stanley. This could be true if you do not mind a howling gale in your face!

Fig. 2-8. A Stanley of the early Twenties with touring body is, from the exterior, virtually indistinguishable from the contemporary gasoline cars. To charge batteries used for electric lighting, there was a dynamo driven from the back axle.

Fig. 2-9. This Foster steam car, circa 1904, is registered in England and carrys the badge of the Veteran Car Club—not necessarily signifying that it has been officially dated by them. As this car has only half elliptic springs, reach bars are not required.

Fig. 2-10. The White steam car was one of the main rivals to the Stanley. This example of 1903 has been partially sectioned to show the water tube "Flash" type boiler and the engine. The valve gear can be clearly seen, showing the engine to be double-acting—that is, the pistons are pushed up by steam, as well as down! This is one of the designs that put the engine in front and used a conventional gasoline-car type propeller shaft and live-axle transmission. The "radiator" is a condenser which greatly increased the range between water refills. The tire shape is not recommended.

from 10 hp for the early cars up to a maximum of 30 hp in 1908, finally ending up with 10, 15, and 20 hp ratings in 1913—all with shaft drive. This was, in fact, the last series production steam car made in England.

That specialist steamer, the Doble, had a four-cylinder compound engine, with the cylinders cast in pairs, two being high pressure of 2.67-inch diameter and two low pressure at 4.5-inch diameter each. The stroke was a 5-inch and the engine was integral with, and geared to, the rear axle. The flash boiler had nearly 600 feet of cold-drawn, seamless steel tubing, coiled into a diameter of 22 inches and a depth of 13 inches and was tested to 7,000 p.s.i., although working pressure did not normally exceed 1000 p.s.i. From cold, the car could be ready to move off in 30 seconds or less and could exceed 100 m.p.h. The condenser was an ordinary radiator, mounted at the front, behind which was a giant fan. The whole car was built to a very high standard, with particular care applied to the braking system.

Chapter 3 Finding Something to Restore

Having briefly discussed the various applications of steam to stationary and road vehicles, if one wishes to restore such, it might be of help to obtain one first!

There are a number of ways of achieving this objective. Probably the easiest is to see what you want and steal it. If, however, you are one of the few remaining mortals who consider it morally wrong to steal, maybe an ancient relative will die leaving his country estate to you and buried therein you may find the steam engine of your dreams.

It is more likely you will be like many collectors and have to acquire the machinery the hard way, by searching for it. It is not much consolation to know that the first vehicle or engine is the most difficult to find. Once you have restored one, you will find that the contacts you make in the process are quite likely to lead you to others you can acquire for future projects. However, this does not help the newcomer. (See Fig. 3-1.)

The first move is to put advertisements in as large a number of magazines and papers as you can afford. The prime U.S. magazines are *Hemmings, Old Cars* and *The Steam Automobile*; plus, for stationary engines, there are the newspapers read by the farming community.

You will find that a large number of farms in existence at the turn of the century will have had a steam engine for running their plant, such as their thrashing machine. This may have been of stationary, or portable, type or, if you are exceptionally lucky, it may have been a self-moving portable or traction engine.

Ferreting Around Farms and Factories

As well as using local papers for ferreting these out, it is well worthwhile spending a few days in your car going around the outlying

Fig. 3-1. The Willans engine of 1888 is a rather curious, though compact, triple expansion unit with central piston valve. This would be a most interesting restoration project, but one you will be lucky to find. Note the three bearings on one crankpin.

farms asking if they have anything in this line. Remember your manners. There is no reason why a farmer should be pleased to see you. Go in quietly and pick a suitable time to start asking if he can help you. Milking or haymaking time is not a good time. The best approach is usually not to ask outright if an engine is on the farm, but to ask if they know of any on any farm in the area. Even if you do locate one, you may well find the owner reluctant to sell it. It is no use telling him he will never restore it, and it would be far better to let you have it. Why should he?

The far better way is to start talking about the old engine, ask what it was used for, say how nice it would be to see some of the old engines working again, and gradually steer the conversation around to the prospect of you restoring it and, when finished, letting the original owner play with it. Not one in 100 of the original owners ever want to play with it, but the gesture often persuades them to part with what may be to them a family heirloom. (See Fig. 3-2.)

Another possible source is any factory that has been established a long while. Steam power was the earliest form of power supply, apart

Fig. 3-2. An example of the Robey semi-portable, over-type engine—hardly portable at all, in fact! Although similar to locomotive design, note that the cylinder is at the firebox end, the reverse of the usual railroad practice.

from the water wheel, for most factories. Most of these engines were rather large, probably too large for most people to entertain; however, some of them might be reasonable. If one of these does interest you, you can make a direct approach to the factory manager. He is usually a busy chap who will not appreciate small talk and will give a straight answer to a straight question.

Having completed your deal, whether it be at a factory or a farm, when you collect, or arrange to have collected, make sure you are there yourself to check that the engine is lifted very carefully. It is all too easy to fracture an ancient casting. Also, have a good look around to see if there are any bits and pieces which might belong to the setup lying adjacent. Often small parts get detached over the years. Also, if there happen to be any old employees about —really old men with long gray beards—ask them if they remember the engine in use and if any alterations were carried out to it, or if there were any regular problems. You may well pick up tips on its operation at the same time and, possibly, even offers of help.

Finding Road Vehicles

Road vehicle acquisition needs a different approach, as it is very unlikely you will find such on a farm or in a factory. Although the odd, undiscovered vehicle does still turn up, it is rather exceptional these days. A more likely source is the old steam car world (See Fig. 3-3.)

Advertisements in car magazines are the first step. Next, join the Steam Automobile Club and go to all the old car meets that you can. Try to become known to steam car people, but do not expect them to be overly friendly. All of us in the old car world are mad. We must be to play with such toys, but those of us involved with steam are eccentric as well! Generally, we are skeptical of newcomers until they have proved their interest and are accepted into the play group. As you gradually get the feel of steam vehicles, you must decide if you are prepared to make do with any type, or if you want a particular make or design. The former is much easier to find than the latter.

Prepare to Build Your Own

When you are on talking terms with one or two steam buffs, ask them to tell you about their vehicles—get to know the good and the bad points. As they gradually accept you as one of them, you will be in a position to ask help in finding a vehicle for you to restore. What you will get will, of course, depend upon what is available, but also on how much you can afford to spend. If money is a problem, be prepared to start with very little and gradually acquire the bits to complete the vehicle. For example, if you covet a Stanley, look out for the engine and rear axle, since these are the basic bits which you virtually must have from the original. The rest can be made or adapted from other parts, if it proves quite impossible to find any more than this. If you have to adapt other parts, make sure you end up

Fig. 3-3. Undercarriage of early Stanley, no chassis proper, but merely reach bars to locate the axles relative to each other on their full elliptic springs. The body could, and did, wobble like a jelly relative to the wheels!

with as near a copy of the original that it fools the experts. Take careful measurements of the parts you need from an original car, and then copy faithfully. You may well be able to adapt running gear from an entirely different make by careful work.

The special steam bits, such as pumps and valves, you should be able to get from a steam man, even if you find them in ones or twos all over the world. Go to all the flea markets you can and search every stall—tedious, but frequently rewarding. Boilers will nearly always have to be made new, or made yourself. See Appendix B for commercial makers.

Bodywork is really not difficult to make, at least for vehicles prior to 1920 which covers the majority of steam vehicles. The construction was usually a wood frame covered in steel or aluminum panels, mostly of simple form. Fenders are also quite easy to make for all pre-1912 vehicles. In the case of cycle-type, they are still manufactured.

This may sound like hard work. It is! It may be considered by some to be cheating. The author does not agree, providing the reconstruction is sufficiently close to the original design that no reasonable differences can be detected and similar materials to the original are used. To those who disagree, we pose one question: What about those supposedly original cars that, in fact, were involved in a serious accident during the first few years of their lives and, as a result, had to have new major components, including maybe the frame, fitted by the factory? Do they then become of later date than their original manufacture? Currently in our workshops is a car less than a year old that was so severely damaged most of the panels have had to be renewed. What does this become? The panels have, in fact, had to be made from new steel, as it was of such limited production that no spare body parts were available.

Dealers and Auctions

If you can afford it, life is certainly very much easier if you can find a complete vehicle to restore. These are, however, rare. As well as the aforementioned methods of location, if money is no object, it is worthwhile trying the steam vehicle specialists and the old car dealers in general.

The other possible source is an auction. These are advertised in *Hemmings* and *Old Cars*, and the ads frequently list some of the vehicles that will be offered which very occasionally include a steamer. However, buying at auction can be fraught with problems. Often it is difficult to tell if you are bidding against another buyer, the owner, or the parrot sitting on the auctioneer's shoulder. If you succeed with a bid, you are virtually saddled with what you have bid for. Guarantees are not worth much for steamers, and it is almost impossible to get other than a superficial idea of its condition from a pre-auction inspection.

Before commencing restoration, read all you can about the subject of steam engines and their equipment and ancillaries. Try to check on the original specification of your particular item, and endeavor to locate other people with the same or similar model.

Remember: Your life may well depend upon how good a job you make of the restoration!

The exhaust of a steamer may be clean in the environmental sense, but in most other ways steamers are dirty, mucky things. Solid fuel is dirty and liquid fuel is smelly or greasy or both! Hot steam pushes out oil and grease with gay abandon, and if you hope to have an operational, immaculate showpiece—forget it. You either do not use it very much, or you use it as originally intended and, ere long, have it all greasy and smelly unless you wish to spend most of your waking hours cleaning! However, hot cylinder oil combined with a whiff of coal smoke is considered by the real enthusiast to be a delightful aroma!

Chapter 4

Three Types of Boilers

There are three main types of boilers:

- Firetube or locomotive
- Water tube
- Monotube or flash

The first two are pressure vessels containing a fair quantity of water at any time and having a capability of an instantaneous increase in steam output for limited periods. (See also Fig. 1-11.) The difference between the two types is in their methods of heat transfer from the source to the water. As the name suggests, the firetube boiler takes the products of combustion through tubes which are surrounded by water, whereas the second type conveys the water in tubes through the heat. (See Fig. 4-1.)

The monotube boiler is, in effect, one continuous tube to which heat is applied externally. Water is pumped into one end and "flashes" into steam in the red-hot tube, emerging as high pressure steam at the other end. Unlike the fire- or water tube types, the flash boiler has no inherent capacity and output is dependent entirely upon the rate of water input. This nil capacity means that the boiler can produce steam very quickly from start up and rates of production can be varied rapidly. These characteristics make this type of boiler or "generator," as it is sometimes called, the most suitable type for private car use.

In considering boilers for restoration the chief parts of the firetube and water tube types are the shells and tube-plates. These are usually of mild steel plate varying in thickness according to the size and designed working pressure of the particular boiler. (See Appendix A for formulas relative to dimensions.) Construction may be by riveting or welding to form a cylinder with ends. Very often the "boiler shell", as it is known, will be covered by a layer of insulating material which is, in turn, encased by a thin sheet of cladding secured by bands. (See Figs. 4-2 and 4-3.)

If a boiler has no current certificate of insurance, all this covering may have to be removed to reveal the shell for inspection.

Once exposed, the boiler can be given a thorough surface cleaning by wire brush, hand or powered, followed by a very close visual inspection. Any covers or plugs should be removed to permit this inspection to cover the internal surfaces as far as possible. Obvious faults to look for are deeply corroded areas, rivet heads burnt or corroded away, and cracks, the presence of which will usually have left tell-tale signs of leakage. Any previous repairs to the boiler should be closely scrutinized.

If you do not mind very hard, dirty work, the fitting of new tubes may be undertaken; but any necessary repairs to the shell should be left strictly to a pressure vessel expert.

Tube Removal and Replacement

Replacing tubes depends upon the construction of the boiler. On firetube types, one usually has to get the tube out from one end or the other, drawing or pushing it right through the boiler. To do this the end of the tube must be drilled away. In expanded types, it is often necessary to drill down the end of the tube for the depth of the tube plate.

First, try drilling just one end and then see if the tube will drive out through the other end. If not, drill both ends; but, before doing so, keep the tube located by means of a bin steel bar in the end drilled first. Otherwise, after drilling the other end, the tube is likely to move out of position and become impossible to remove. To assist in driving out tubes, it helps to make a pair of stepped mandrels, one to locate the end being pushed out and the other to use as a drift.

When ordering new steel tubes, they should be made of cold-drawn, seamless mild steel in the best quality you can get. Do not forget to allow in length for any protrusion that is needed beyond the tube plates. Some tubes are swaged up in diameter at the smoke box end; in which case, mention this fact to the steel supplier. Wall thickness depends on the size of boiler and its design, but is usually within the range of 8 to 16 gauge, or from a little over 1/8-inch to a little under 3/32-inch.

Ideally, the new tubes should just slide into place with a wooden mallet. However, before attempting to fit them, anneal the ends. In the case of copper anneal by heating to a cherry red and quenching in water. In the case of steel, anneal by heating to a dull red and allow to cool slowly in a reasonably warm atmosphere. This is necessary to enable the tubes to be expanded into the tube plates.

The holes in the latter must be carefully scraped clean. An old file is ideal for this, finishing off with emery paper to get a bright surface.

If the new tubes are slightly oversized, they can be reduced on a lathe, but no more than ten percent of the wall thickness may be removed in this way. Do not attempt to do this with a file, unless you are an expert

Fig. 4-1. This Galloway boiler is in the upper end of the size range, probably larger than most restorers will encounter. Nevertheless, it gives an idea of the multitude of boiler designs that exist. It has two massive firetubes running throughout its length, with quite large water tubes criss-crossing these firetubes all along. Although basically of "Firetube" type, it does have elements of the water tube design. Safety valve and steam take-off are obvious on the top. The lower cock is the drain, or sediment, valve.

Fig. 4-2. A typical small firetube boiler used for plants up to about ten horsepower. This example has, in fact, got copper firetubes, each about ½-inch diameter. If you want to know the number, count them! Shell size is about 17-inches diameter by 15-inches high, constructed of steel with steel tube plates. The base shown is merely a stand and not part of the boiler, but something of this general shape could be used for a solid-fired stationary boiler. The level gauge on the side is not, repeat not, the type we ever recommend you use. We consider that no matter how thick the glass, this type of gauge is for low pressure use only. Steam take-off is in the center of the top tube plate.

metal craftsman, otherwise you will end up with multisided, rather than cylindrical, tubes that can never be made to seat properly.

Using a Tube Expander

Some people prefer to fit all tubes and then start expanding them at each end. Others prefer to fit and expand each tube, one by one. The choice is yours.

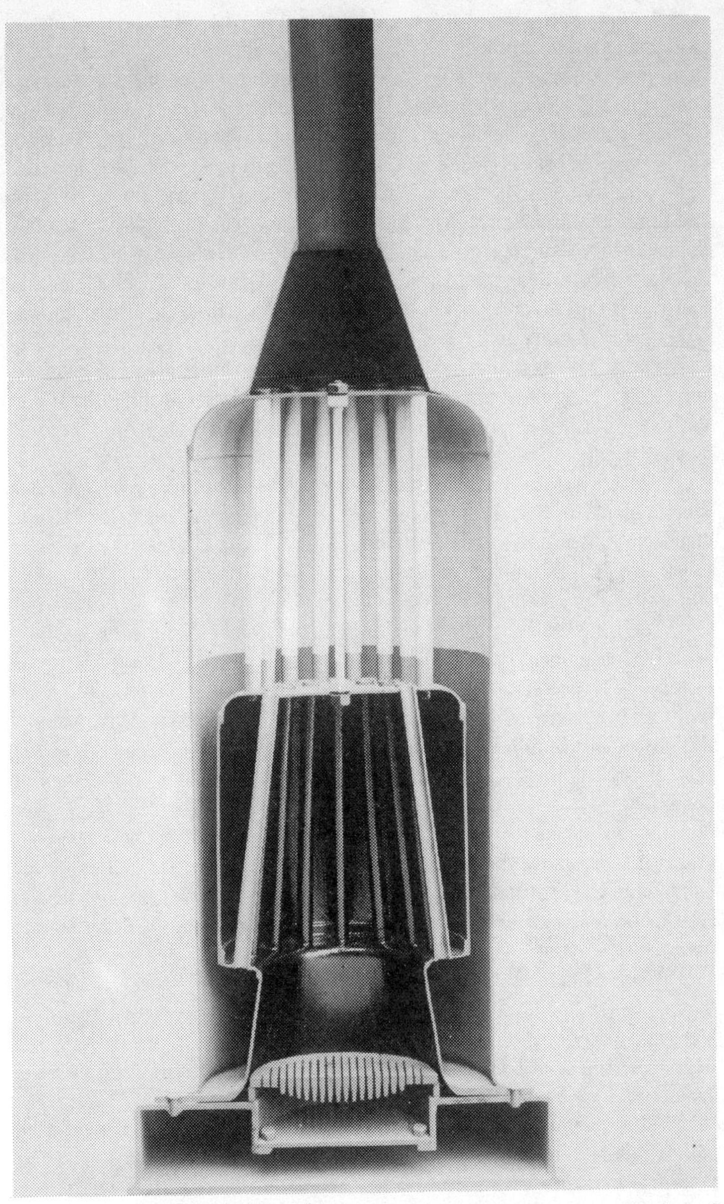

Fig. 4-3. The Richardson vertical boiler is interesting as it is a combination of water tube in its lower half and firetube in the top half of its construction. Again, it is a solid fuel-type burner with the grate and ashpan forming an integral part of the unit. This design is normally used only for stationary engines. Note the screwed rod traversing the firetube half. This is known as a "stay," and is to prevent the tube plates from bowing or buckling under steam pressure.

A tube expander is basically a taper pin forcing rollers out sideways onto the internal surfaces of the tubes, thus stretching or "expanding" them into the tube-plate holes. If you cannot borrow or purchase such a device, it is not difficult to make. The most tedious part to make is the cage to hold the rollers. This is akin to the cage of a roller bearing and is most easily made in brass. In effect, it is a bush to fit loosely inside the tubes with three slots, 120 degrees apart, to hold three rollers about one and a half times in length and the thickness of the tube plates. Having then found or made three suitable rollers—these need to be hardened steel if the tubes are steel—a tapered central pin must be turned up of such diameters that, when pushed down inside against the rollers, it pushes the latter out tight against the tubes.

To use the expander, steady and fairly substantial pressure must be applied on the central pin, or mandrel, as it is turned. Shaping the top in the form of a hexagon enables the normal socket wrench sets and their various handles to be used.

The only guide that can be given to the amount of expanding each tube requires is that usually one can feel considerable resistance to the mandrel being turned when the tubes are tight. Do not go on so long that the tubes are split, as it is better to err on the slack side and retighten after a cold water test, if they leak. (See Fig. 4-4.)

Keep the expanding tool well oiled and work around the tubes in a progressive sequence, not at random. If your boiler has the stepped-type of tube, with the smoke box end slightly larger, you will probably need two mandrels or, if you are lucky, just two sets of rollers. Make sure in this case that the rollers are only running around the increased end of the tube and not going in so far that they are in part running on the smaller diameter.

In some cases this is the only method of keeping the tubes tight, but other designs allow the tubes to protrude beyond the end plates. After expanding, the protrusion is beaded—that is, flanged over flat against the tube plate. This not only helps to eliminate leakage, but also gives some measure of staying strength to the boiler. (See Fig. 4-2.)

Another addition is the insertion of ferrules. These are tapered bushes which are driven into each tube end, after it has been expanded, and which, in effect, increase and tighten the expanding process. When making ferrules, radius the corner to prevent them from biting into the tubes and allow a taper of about one in 25.

The advice of your insurance company's boiler engineer should be sought before commencing any repair as, ultimately, he has to declare the boiler fit for service. Unless carried out within the last five years, the boiler should be subjected to a static cold-water pressure test to at least one and a half times the normal working pressure and held at that for 20 to 30 minutes. This process will show up any major defects such as deformation of plates or leaks. Small leaks from around the tubes can be discounted as these will usually disappear at normal working temperature and pressure.

Fig. 4-4. The Cochran boiler shows fairly clearly the tube ends expanded into the tube plates and is of quite rigid rivetted construction, not needing any stays. This solid-fueled vertical boiler has water passages close around the fire, as well as firetubes, through the water space above the grate. Note the balance weight safety valve.

Winding Wire-Wound Boilers

A variation on the firetube design used by the Stanley twins and one or two other people was a relatively thin shell wound with piano wire.

If the wire is the slightest suspect, for instance having anything more than surface rust or showing any signs of breakage, or it is known to be the original and somewhat ancient, remove and replace it with new wire, 0.054-inch diameter of 325,000 pounds per square inch breaking strain.

It is not a particularly easy job to rewind as many thousands of yards of wire as are needed to close wind three layers, if the boiler is 20-inch diameter or smaller, and four layers if up to 25 inch diameter. Above this size, wire-wound boilers are not normally used. The only sensible way to rewind this type of boiler is to have some means of mechanically turning the boiler shell, such as a large lathe, and then, when it is turning quite slowly, to gradually feed the wire onto it with each rotation being as close to the previous coil as possible.

When the original wire has been removed, check the shell for any signs of damage or bulging. Usually the end plates can be retained.

Boiler Cladding

The cladding should not be replaced until the cold-water test has been completed, and then new material should be used, if it can be obtained, unless the original is in very good condition. Often this original was asbestos of some sort or other, and although these days it is fashionable to decry asbestos as dangerous, the small amount involved in this operation is of minimal risk. It is not nearly as great a risk to life and limb as the general danger of steam engines!

There really is not any more easily used substance for this cladding than asbestos, but if you cannot get it, one of the various types of fiberglas now available can be used. Try to find one that is fire resistant.

If you are entering competitions with your steam apparatus, you will lose points for using fiberglas cladding, if you are found out!

Do not forget that the boiler is the most dangerous part of any steam equipment. Whatever you may do on other parts of the system, do not ever take chances with the boiler, no matter how tempting it may be.

Water Tube Boilers

Most of the preceding information applies equally to water tube boilers as these are a sort of inside-out version of the aforementioned tea kettle and, as the name implies, have the water contained in tubes exposed to the products of combustion.

There are many detailed variations of the type, but generally two large diameter tubes or drums are connected by numerous small tubes arranged, to borrow an electrical term, in parallel. (See Fig. 4-5.) The tubes themselves are seldom straight, often being in a "U" or "V" configuration arranged horizontally or vertically. This is because the

Fig. 4-5. The Gurney water tube boiler is of early nineteeth century design, with the firing doors between steam and water drums. The hot combustion gases are drawn over the tubes and deflected up and down by the brick baffles until they reach the chimney. The safety valve is of the balance weight type.

water tubes are capable of operating at a much higher temperature than firetubes and, thus, are subject to considerably more expansion. The design bend allows for this expansion to take place without over-straining the attachment of the tubes to the drums. The common chamber connecting the tubes at the lower part of the boiler is known as the "water drum," while its counterpart higher up is the "steam drum." The tubes are fitted in the drums by expanding as described for firetubes. (See Fig. 4-6.)

Certain designs of boilers use short, straight water tubes either in association with firetubes between tube plates or arranged in a grid formation. The latter type is often all welded, as the diameter of the drums precludes access by an expander. (See Fig. 4-7.)

Examination of the boiler starts by withdrawing the drums and tube assembly from its casing, disconnecting any pipes or fittings necessary to achieve this. A visual examination will reveal any obvious faults, such as burning and pitting. Remember that you are examining the fire side of the device! Internal examination of the water spaces is usually limited to the drums via hand or mud holes, though some boilers with straight tubes have screwed plugs opposite the ends.

Should a tube be found to be burned out completely, the most likely cause will be an accumulation of sediment due to impurities in the feed water. This forms an insulating coating on the walls of the tube, thus depriving it of the cooling action of the water. Cooling water at 350 degrees F. may raise eyebrows; but, remember, the alternative is the flame at, say, 1000 degrees F. plus. If a tube, or tubes, is found to be burned, then you must consider the whole lot as suspect since they have all had the same diet.

Replacing Tubes and Insurance

Equally, deep pitting due to acidic products of combustion is sure to have affected more than the odd one or two. No hard and fast rules can be applied as to the sort of tube used for replacement, as this depends upon

Fig. 4-6. This Stirling boiler is an example of the water tube type with a solid fuel burner grate built integral with it. Note the steam drums at the top and water drums at the bottom. Steam take-off is in the center drum, and the small diameter pipe leading from the right-hand drum goes to the steam pressure gauge. There are balancing pipes between drums.

the design of the boiler. Generally, use the same diameter and wall thickness as the originals. This may be ascertained from a sound piece after the tubes have been removed. The material is usually steel, although some designs employed copper. When obtaining replacement material, solid drawn is preferable, although electric resistance-welded tube may be permissible. *Do not be tempted to use common water pipe or any galvanized finish.* The first is not suitable at the temperatures encountered and the second, while seeming a good idea, will burn and flake off internally.

If in doubt, consult the company which will ultimately be insuring the boiler and you in the event of the thing "going bang." Although it may not necessarily be compulsory, do not be tempted to do without boiler insur-

Fig. 4-7. The Perkins water tube boiler of the 1870's has the rather unusual construction of threaded hollow tube connectors between adjacent water tubes. It is fairly easy to make and repair and has the ability of being made to, more or less, any required capacity. Note the substantial protected water level gauge on the right-hand side; above it is the steam pressure gauge.

ance, no matter how difficult you may find it to obtain. It only needs one serious accident involving a third party and a lack of compensation to put a stop to what is an absorbing hobby for hundreds of people.

The drums must be treated rather like the pressure shell of a locomotive-type boiler. Look for corrosion and pitting internally and externally, wastage (thinning) in localized areas, and cracking around flanged parts and rivets. Should these faults be absent, proceed to the attachment of fittings, examining threads, studs and mating faces. Parts screwed directly into a boiler are usually plugs of one sort or another, or studs by which other parts are attached.

Renewing Threads and Plugs

These threads must be perfectly sound and exist for the full depth of the plate. Should this not be the case, there are several possible cures. In the case of a damaged thread, it may be possible to clean out the hole and cut a new thread of a size larger. In the case of a plug, a new one must be made to the same taper as the thread employed, noting that when in tightly there must be several threads showing on the plug. This allows the plug to screw in tightly at subsequent removals without the danger of less than full thread depth contact. If a threaded stud needs to be replaced with a larger size, the new stud may be made with two diameters, the larger being threaded oversize to the original to screw into the enlarged hole that has been retapped, and the smaller original size threaded the same as the previous stud.

This is probably the best cure for threads which are in a pad, or thickened part, forming part of the boiler plate. The pad idea can be applied to reinforce a short, but otherwise undamaged thread by welding on a plate externally and by drilling and threading, thus, giving substantially more length of thread. When these cures are not feasible, the only real answer is to cut out a piece of the boiler plate containing the offending thread, weld in a new piece, and drill and tap to the original thread in the appropriate place.

Unless you are very skilled in the art of electric welding, this process should definitely be left to an expert.

Mating surfaces are usually provided by the aforementioned pads, which should be truely flat—this state being most readily obtained by judicious application of a good file.

A variation of the water tube type is the aptly named thimble tube. This consists of blind end tubes projecting into the heat but not actually going anywhere. The tapered section promotes internal circulation of the water and the fitting of these tubes in a plate greatly increases the surface area exposed to the heat. The tubes are expanded into the tube plate in the usual way. (See Fig. 4-8.)

Before moving on to consider the last type of boiler, there is a hybrid that is worthy of mention. This is a quite successful marriage of the fire- and water tube in one boiler. Whether such efforts and complication are

Fig. 4-8. The Clarkson boiler depicted is an example of the thimble tube type of the early Twenties, with a section of the thimble being shown at the base. This was one attempt to get increased heating area and, although not that common a design, close inspection does give an idea of how the tubes or thimbles are expanded. This particular example employs waste heat.

worth the theoretical advantage of extra heat extraction is debatable, although there may be few of these boilers left.

Flash Boilers

The third type of boiler listed in the heading of this chapter is the so called "Flash" or monotube. It is really a water tube type and is particularly popular in many steam cars, especially the light-buggy type. In this there are no drums or containers of any volume of water, but the boiler, or "generator" as it is sometimes termed, is just one long tube coiled or turned back on itself many times in the combustion space. It is extremely simple in action, the water being pumped in at one end, tepid or cold, heated rapidly by the fire surrounding the tube and emerging at the other end as steam. The fact that it has virtually no capacity means it avoids the stringent standards applied to pressure vessels. However, it has the disadvantage of a time lag between demand and supply of steam when, for example, sudden acceleration is needed. (See also Fig. 2-3 and 2-10.)

Not much can be done in the way of repair to this type of boiler. It is much simpler to renew the whole coil, again using tubing of the same diameter and wall thickness as the original.

Due to the comparative strength of small tubing, these boilers often work at quite high pressures (900 to 1000 p.s.i. are common). Thus, for a given power output, the use of pressure in this region makes for quite a compact unit. The output end of the coil is usually continued into another coil exposed to the hottest part of the flame in order to superheat the steam, thereby cutting down condensation losses in the engine. Between the water pump and the generating coil proper, the pipe may be led around the inside of the boiler casing, thus preheating the contents by using otherwise virtually wasted heat.

Connections in the pipe between feed, generator and superheater coils should be by a ground-finish, coned union in order to stand the temperature and pressure. The pumps and controls for this type of boiler are dealt with in a later chapter.

Casings for Boilers

The casings for water tube boilers are usually sheet steel of simple construction lined internally with some form of insulating material. On stationary plants, this takes the form of fire bricks or molded fireclay; but, due to the inherent vibration, soft asbestos sheet was almost invariably used originally on mobile outfits. The most likely fault to be found with the casing is rust, ranging from surface attack to complete surrender by the casing to the dreaded tin-worm (a voracious little beast!).

In this case, the answer is to make a new one. The old casing or its remains can be used as a pattern and suitable sheet cut and formed to shape, then riveted or welded. Fire cement mixed to a stiff sort of dough can be applied by trowel and wet hand to the interior to form the lagging. A few internal projections on the casing are useful to "key" the clay in place. Included in these will be the various brackets which support the coils in place.

From the foregoing it will be appreciated that there is a lot the determined enthusiast can undertake himself on boiler repairs, if he is careful.

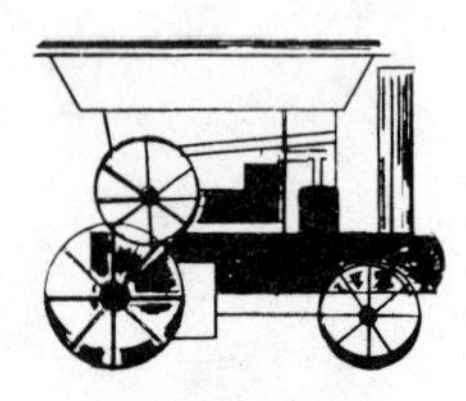

Chapter 5
Boiler Safety Devices

Having gotten fuel to your burner and water to your boiler, if you light your burner, you might be lucky enough to make steam in your boiler! You might just be so fortunate that you can make more steam than you need, even reaching maximum working pressure. To go much above such pressure is unwise as, if you do, it is extremely probable that a rather large cloud of steam will lift you high above to clouds even higher, before depositing you back below the earth where, doubtless, your sins will ensure you remain!

Safety Valves

To try to keep you in this world, a thing called a safety valve is used. (See Fig. 5-1.) This is in effect, a spring-loaded valve, sometimes adjustable and sometimes not, and set to blow off when the steam pressure exceeds normal working pressure (N.W.P.). The overhaul of this is either impossible—that is, it is sealed—or very simple, merely by replacing the spring, if you have any doubts as to its condition, and facing the valve and seating. (See also Fig. 4-1.)

Fortunately, it is a fail-safe device, in that any wear or damage is likely to make it blow off early—that is, at too low a pressure.

All boilers should be fitted with two safety valves, the aggregate minimum area of which—whether on coal- or oil-fired boilers and whether working with natural or forced draft—can be found by the formula quoted in Appendix A.

If the steam is led from the safety valves through a waste pipe, then this pipe and any passages leading to it must be not less than 1.1 times (in cross sectional area) the combined area of the valves, as determined from the formula.

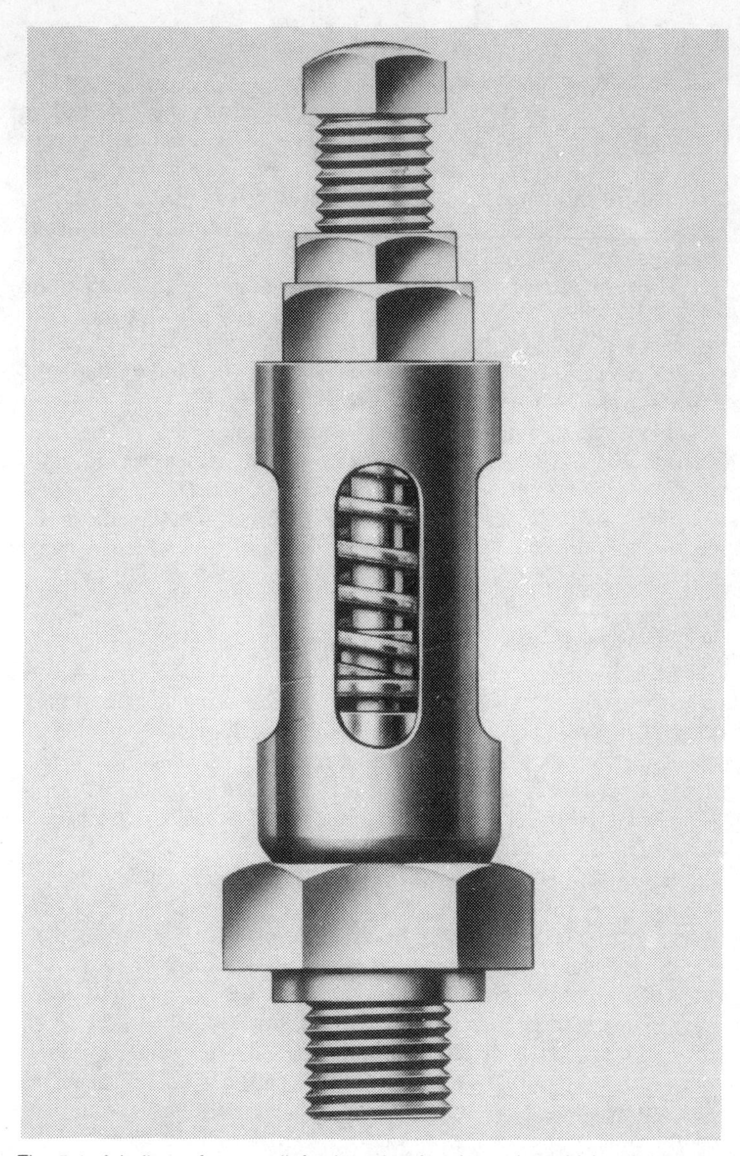

Fig. 5-1. A boiler safety or relief valve, showing the spring which can have its pressure adjusted by the setscrew and lock nut at the top. Do not use this as a means of increasing working pressure above the safe limit for your boiler.

The most usual type of valve encountered is one incorporating a compression spring acting on a wing-type valve. One slight disadvantage of this type is that as the valve lifts, the spring is further compressed, thus, increasing the loading; but, this does not seem to be any problem in operational use. Probably the most common type used in recent years is the Ross "Pop" valve. This is the patent of R.L. Ross & Co. of Stockport, near Manchester, and incorporates the features of very prompt opening and closing and large steam escape area from the immediate surroundings of the valve face. Adjustment is made by either screwing down the upper part of the body containing the spring or by means of a projecting screw provided for the purpose. In all cases, provision is made to seal the adjuster so that once set at the desired pressure, no further tampering by the operator is possible.

The springs must be cased in so that they cannot be overloaded, and in the unlikely event of a failure the bits thus released are contained rather than trying to emulate the trajectory of a bullet.

To calculate the sizes of replacement coil or spiral springs, the formula given in Appendix A may be used.

Another item in the array of boiler protection devices is the boiler pressure gauge. The actual gauge is dealt with in another chapter, but it should be mentioned here that it is a good idea and, in some cases, compulsory for the gauge to have the normal working pressure marked on the face of the dial with a red line. The gauge should have total graduations to at least twice working pressure.

Boiler Water Level Gauges

All boilers, with the exception of flash or monotube types having virtually no capacity, should, if at all possible without spoiling the original design, have two independent means of indicating the water level therein. In fact, the authors would go so far as to say that if the boiler is solid-fuel fired, it must have a second gauge added if not so fitted originally. If it is liquid-fuel fired and has only one water level gauge, it must have a low-water, automatic fuel cut-off valve.

It is usually found that, in practice, most boilers (except those in road buggies) have two glass-type gauges of some sort. Where these are not fitted in a position to be readily visible by the operator, as for example a water tube boiler with the header drums high up, a prism or mirror system with a light may be employed so the level is visible at the operator's level.

There are three main types of glass gauges used. They will be dealt with, in turn, and it must be stressed that the safety of the boiler, yourself, and any other persons in the vicinity can depend upon the reliability of the water level indicator. Probably more boiler explosions have been caused by running low on water than through any other cause.

Tubular Glass. The tubular-type consists of a glass tube mounted vertically in two fittings projecting from the boiler shell. (See Fig. 5-2.) These fittings may be screwed directly into the shell in smaller boilers, but are more usually flange mounted and secured by studs and nuts.

The threads of the former type should be in perfect condition and annointed with a jointing compound when screwing home. The flange mountings should have a gasket of a suitable, thin (1/32 inch) steam jointing

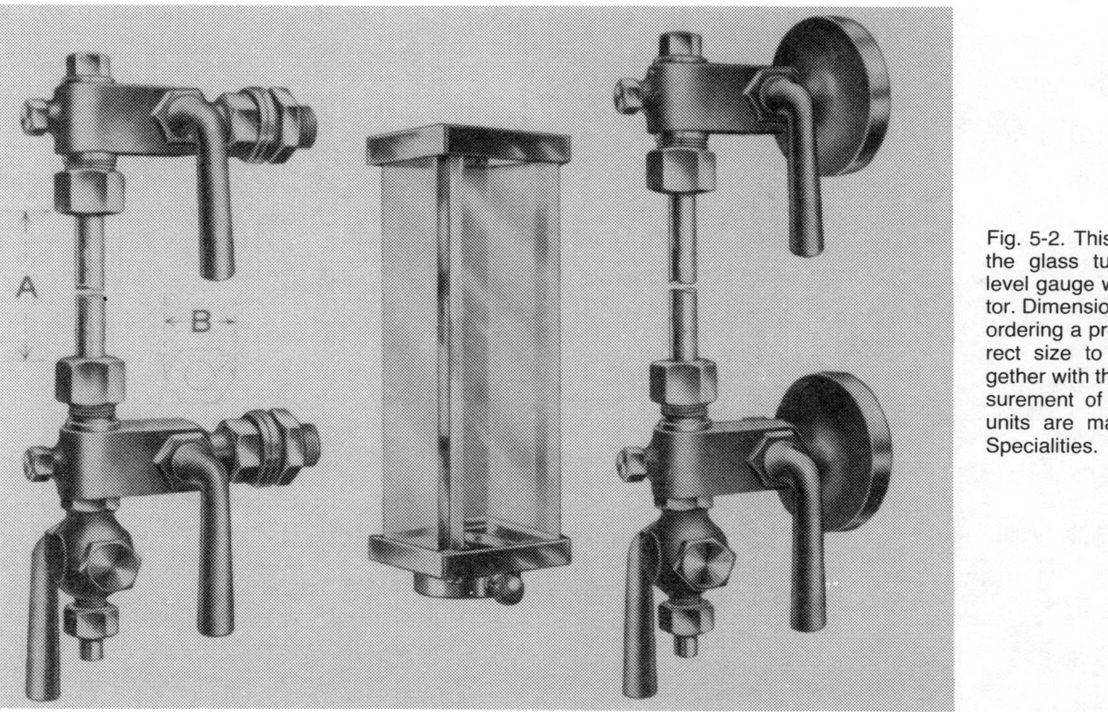

Fig. 5-2. This is a clear example of the glass tube type boiler water level gauge with stout glass protector. Dimension "B" is required when ordering a protector to give the correct size to the manufacturer together with the obvious length measurement of "A." These particular units are made by British Steam Specialities.

material. When fitting either type to the boiler, great care must be exercised to ensure that the bores that will take the gauge glass tube are perfectly in line, vertically when viewed from both the front and from the side. This latter consideration is achieved by inserting thin washers to pack one of the fittings further from the boiler.

A steel rod of the same diameter as the glass to be ultimately fitted can be used to check the alignment of the fittings. It is most important that the glass tube is not subjected to any stress from the fittings it has more than enough stress with which to cope from its occupational hazards—one end being connected to water in the boiler, the other end to steam).

Many injuries have, in the past, been caused by the bursting of the glass gauge tube, but there are now two ways of combating this risk. The first is use of automatic shut-off valves incorporated in both top and bottom connections, these being in the form of either flat phosphor-bronze springy flaps or ball bearings. In the event of a glass bursting, the sudden rush of steam and water trying to escape from the fittings causes the flaps or balls to be pressed over the outlets therein, very quickly sealing these off.

To contain the fragments of tube when it initially breaks, a "protector" should be fitted. This consists of thick, toughened glass plates mounted in a heavy frame which surrounds three sides of the tube. The fourth, or open, side is facing the boiler and is sometimes formed of a perforated metal sheet. The protector is easily removable to enable the gauge tube to be kept clean, one knurled screw usually serving to clamp the device to the bottom gauge fitting.

It will be noted that the top and bottom gauge fittings should also incorporate shut-off cocks and, in addition, a drain cock should be fitted underneath the tube and in communication with it. (See Fig. 5-3.) Then, if the boiler is in steam and the gauge needs cleaning, the cocks can be closed in both upper and lower fittings and the drain cock opened to drain the contents of the tube. Leave the protector on until you are sure from opening the drain cock that both shut-off valves are closed.

Internal cleanliness of the gauge is extremely important, as the water passages are subjected to the same impurities that may be present in the boiler. Thus, it is possible that scale or other foreign bodies may be deposited in the passages which will undoubtedly lead to false readings. To guard against this the gauge should be "blown-through" at regular intervals, the procedure being:

Keeping the protector in position:

- Close lower fitting shut valve
- Open and close drain cock
- Open lower fitting cock (shut valve)
- Close upper fitting cock
- Open and close drain cock
- Open upper fitting cock

Fig. 5-3. The glass tube type boiler water gauge—assembled complete with protector glass and lower drain cock. Handles are shown in their normal operating, downward, position.

Additionally, plugs are located in the fittings opposite the waterways to enable these to be periodically examined and physically cleaned. There is a further plug in the top fitting which enables the glass tube to be inserted. The glass tube itself is of a special heat resistant quality and is supplied by the manufacturer to the correct length to suit the gauge. The diameter of the tube is usually in the range of 9/16-inch to ¾-inch, is located in the fittings in a fairly close bore and is sealed with glands top and bottom. These may, in older examples, be packed with graphited asbestos string (a fiddly process); but, more usually these days, a molded rubber ring is used. These are supplied by the gauge manufacturer. Great care must be taken not to overtighten the gland nuts, as this imposes a strain on the glass that could lead to its fracture.

Any corrosive matter in the feed water will cause deterioration of the glass and this should be changed before failure occurs. Signs can sometimes be seen that the glass is nearing the end of its life—notably streaks and flaws at the top end.

Reflex Gauges. The next type, the reflex-type gauge, consists of a heavily constructed, usually cast, three-sided box with the fourth side that faces the operator being a thick plate of glass retained by a frame securely bolted to the main body. (See Fig. 5-4.) The side of this glass which is adjacent to the water and steam is usually formed with a series of closely spaced, vertical "V" shaped grooves. Due to the differing refractions of water and steam, the level is very clearly defined. The space occupied by the water appears black against the silver of the steam.

This type of gauge is immune from broken or burst glasses and, therefore, requires no protector. The thickness of glass is such that even if accidentally struck with a metal object such as a fire iron, the worst that will happen is a crack will appear. The top and bottom fittings are similar to the previously described gauge; but, instead of glands, the reflex gauge unit is attached between the fittings with union nuts. This means that the gauge may be readily turned to face any desired direction for easy reading just by slackening these nuts, turning and retightening.

The third type is really a variation on the reflex design. Its construction is of a similar type, being a heavy casting bolted together, but having two thick glasses on two opposite sides. The water level therein is seen by light shining right through the gauge, which must, therefore, be positioned so that neither of the glass sides is masked and so that the operator can see through it either directly or by means of a mirror. (See Fig. 5-5.)

For any of these gauges, if the plant is used at night, some form of illumination must be provided.

If only one water gauge is fitted, the alternative method of determining the water level is by test cocks. These are, as far as possible, fitted directly to the boiler shell—one at the lowest permissible water level and one in the steam space. If the boiler is more than 7 1/2 feet in diameter,

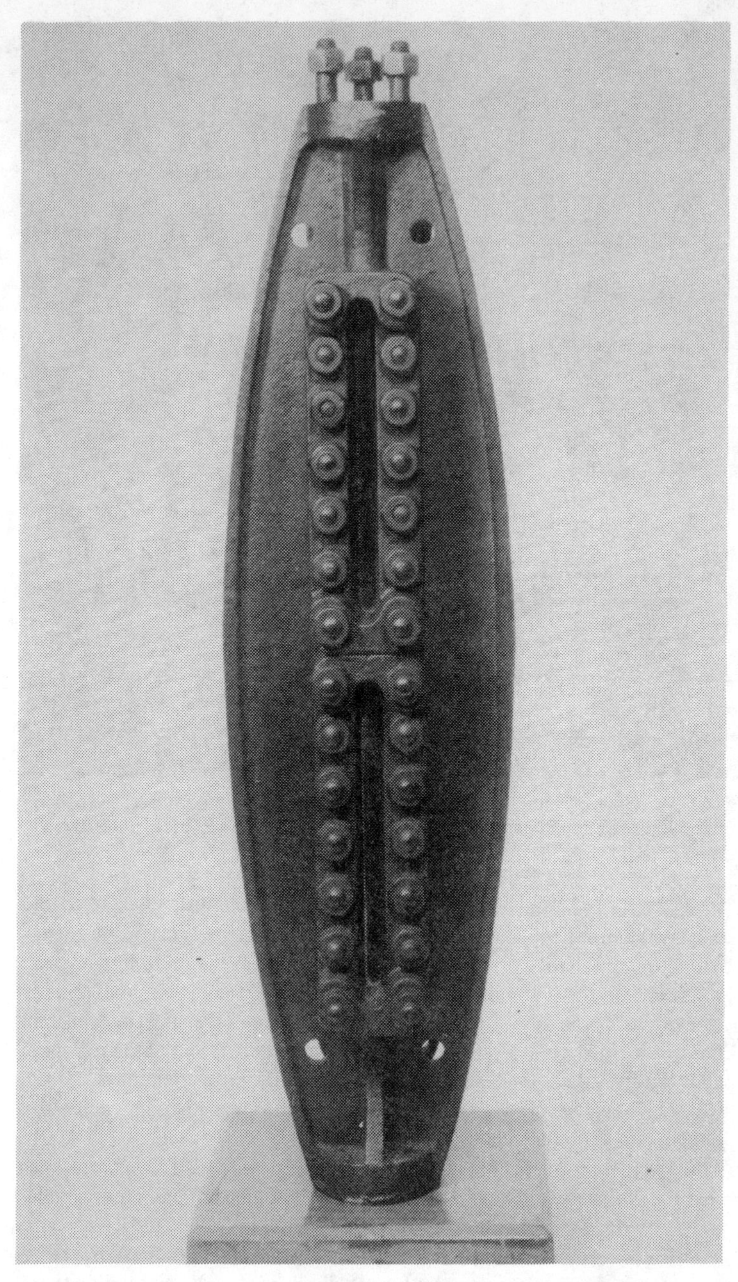

Fig. 5-4. This boiler water level gauge is an early example of the reflex type, being built by Perkins during the latter part of the nineteenth century. It gives a good idea of the heavy construction needed for safety in this application.

Fig. 5-5. View of some of the accessories of an early steam buggy showing substantial cast steel bodied boiler water gauge—It may not look pretty, but it is safe!

an additional cock should be fitted. Under normal conditions, steam should emerge from the top cock, and water from the bottom cock, when each is opened momentarily. If steam should emerge from the bottom cock, the water level is too low and steps must be taken to remedy this situation very quickly. With a liquid-fired burner, it is well to shut off the main jet feed until the water situation has been corrected. If water comes out of the top cock, shut the throttle at once and drain off some of the boiler contents, via the blow-down valve; otherwise, there is a risk of water being taken into the engine and badly damaging it.

Regulation of the fire or burner and the water feed should be such that the demand of the engine for steam is satisfied while keeping a stable, or relatively constant, water level.

Fusible Plugs. However, there is one last line of defense if all else has gone wrong, this being the fusible plug. In a shell-type boiler these are fitted in what is often termed the "roof" of the furnace or, depending on design, the lower tube plate of the boiler—in other words, in the part of the boiler nearest to the hottest part of the fire. Should the water level fall so low as to expose this surface, the core of the plug will melt and allow an orifice open to the boiler through which will issue, at some speed and force, a jet of steam and whatever water is left in the boiler, directed to the fire, which should extinguish it before major damage can occur.

The plug itself can be likened from its appearance to a threaded bush made of brass with the bore filled with lead. This lead will melt as soon as it is no longer covered by water. Some designs have a slight variation in that there is a brass button in the center retained by the lead. The plug or plugs can be unscrewed from outside the boiler, that is on the fire side. Usually the thread is of tapered form and, thus, self locking. Any remnants of lead should be melted out and the plug thoroughly cleaned up. Examine the plug for signs of burning and thread damage, and if in any doubt, replace it.

When making a new plug, it is vital that the thread form and taper should match those in the threaded hole in the boiler. If not, there is the possibility that there will be only one or two threads holding, with consequent danger that the whole plug may blow out. As the mean outside diameter of such plugs may be 1-1/4-inch, compared to a 1/2-inch normal center bore, the loss of a plug will result in such a jet of steam being released into the firebox that damage and injury are almost certain to be caused.

The bore of the plug, which may be threaded to give a better key to the lead, should be tinned and, while still hot, filled with molten lead in one pour. The plug is placed upright on a piece of wood which will retain the lead without getting stuck to it. Even if plugs are not melted during service, they should all be removed and re-leaded annually, as the lead tends to deteriorate where exposed to the furnace. When refitting the plug or plugs, a smear of graphite should be applied to the threads to aid future removal.

Take care not to over-tighten the plug, as this may tear the thread—particularly if fitted in a copper plate. There is one other water level indicating device which you may encounter, especially in some Stanely cars. It has been left to the end of this list of safety devices as it is one that in the author's opinion is not to be recommended. In fact, we would never use one, even if it was original.

Standpipe and Glass Tubes. The device consists of three parallel tubes, brazed together and mounted vertically alongside the boiler. The middle tube mimics the water level in the boiler, its lower end being connected to the bottom and the top to the top of the boiler. Thus, the water level in this tube is the same as that in the boiler. The left-hand tube is part of the pipe leading from the water pump to the boiler. Thus, when the pump is pumping, water constantly runs through it. The right-hand tube is a standpipe, closed at its upper end, and at the bottom connected to a glass water gauge in front of the driver within easy view.

This standpipe and glass tube, with associated pipework, form a U-tube and is filled with water, the level of which when cold is about an inch above the bottom of the glass. If the water level in the boiler and, thus, in the center tube, is above the top of the standpipe, the cold water passing through the left-hand tube will keep the standpipe relatively cool and the water in the glass will remain at the level mentioned above; but, if the water in the boiler falls below the top of the standpipe, the increased heat in the center tube will vaporize some of the standpipe water and the pressure thus caused will push the water in the gauge glass higher, showing that the water level in the boiler has fallen.

It will thus be obvious—maybe!—that when the water is high in the glass, it is low in the boiler and vice versa. Also, the glass only attempts to give an accurate reading when the plant is running. When the boiler is cold, the water in the gauge glass will be at the bottom whether the boiler is full or empty. A false reading of the glass may also occur from the heating-up of the indicator body when the car is left standing with a head of steam. This may make the water rise in the glass giving a false indication that the boiler level is low.

After this description, it is probably unnecessary to elaborate on the reasons why the authors suggest that you throw such devices as far as you can!

Steam Automatic Valves

A liquid-fueled fire is much easier to control than any form of solid fuel. The usual system is to have a pilot light that burns all the time and a main flame that cuts in and out, dependent on steam pressure and the requirements of the boiler and the plant that is being driven.

A relatively simple valve achieves this effect, usually termed a "steam automatic." Its purpose is to shut off the fuel when a predetermined steam pressure is reached. (See Fig. 5-6.) One type consists of two basic units coupled together, one being the steam bit, the other the fuel valve. The steam half contains a diaphragm to which is connected a rod. Steam pressure on one side of this diaphragm moves the rod, the other end of which enters the second half of the unit and operates a valve to shut off the fuel when the desired pressure is reached at the steam end. (See also Figs. 2-5 and 12-1.)

Ideally, at this steam pressure the valve should snap shut. Unfortunately, most early designs did nothing of the sort, in that the rod merely acted on a flap valve or just pushed a ball bearing across an orifice, thus shutting off the fuel supply. This means that it slowly closed as the steam pressure built up, the diaphragm started to move long before normal working pressure, and thus, the fuel pressure at the jets gradually reduced to near cut-off point and, hence, it was quite likely to light back at the jets. If originality does not bother you, you may well feel it worthwhile redesigning this device so that it snaps the fuel supply shut suddenly, rather in the manner of most 1920's light switches of the electric domestic kind.

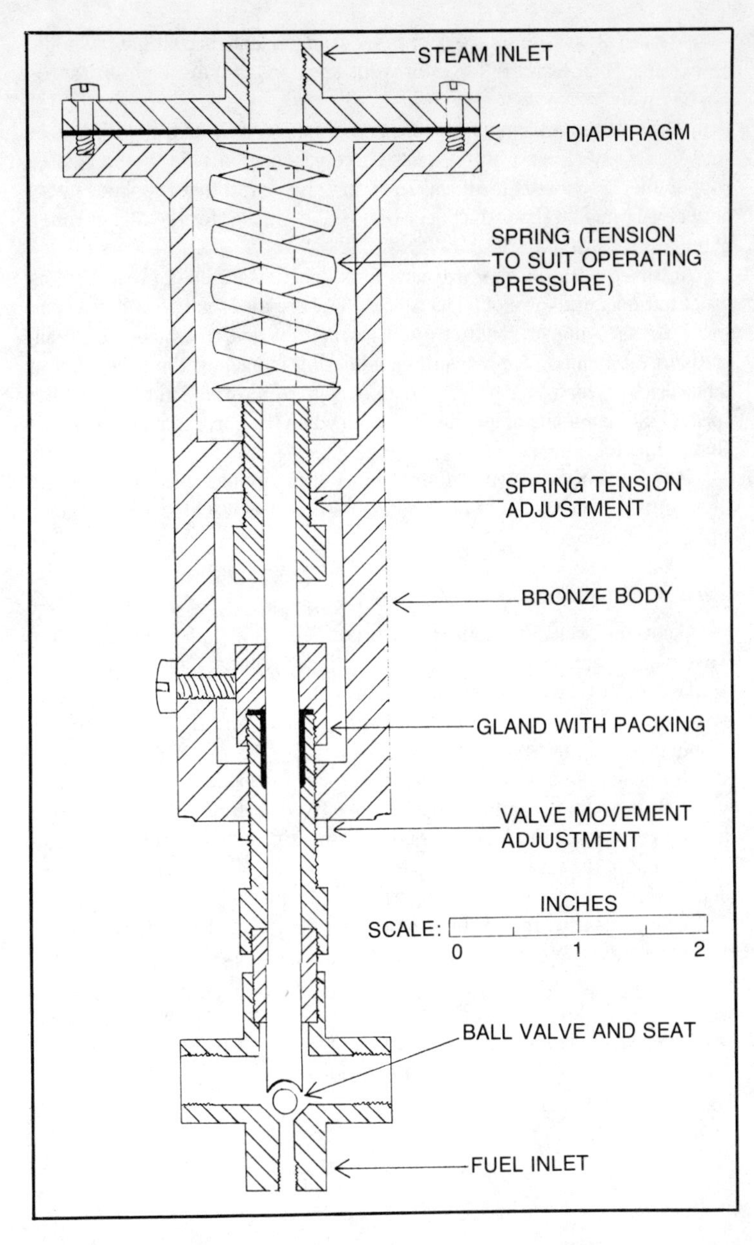

Fig. 5-6. A common type of steam automatic—the device that shuts off liquid fuel to the main burner when the operating pressure is reached—is occasionally found with stationary plants, but more often in steam cars. The diaphragm can be made of hard copper or brass shim stock. The operating rod should be stainless steel.

Whatever system is used, it is most important that all joints should be carefully made and tight, to prevent leakage, as quite high pressures can be involved.

The diaphragm cannot be repaired if damaged or brittle with age or suspect in any other way. It must be replaced by a new well-annealed copper sheet, similar in thickness to the original, or, if that is unknown, try between 10 and 20 thou. thickness or possibly thicker for the higher range of boiler pressures.

In order to alter the pressure at which these valves operate, it is usual to find some form of adjustment. They are made with a compression spring on the opposite side of the diaphragm of that to which the steam pressure is applied. A screwed bush bearing is against the other end of this spring, which can be moved in or out to alter the tension on the spring. Some means of locking this bush when the correct adjustment has been achieved is provided.

Apart from the spring, if steel is used in any part of this valve and replacement is needed, try to use stainless steel as it will last a lot longer without corroding.

Low Water Automatics

Another device sometimes incorporated as a boiler protector for those plants using liquid fuel, is the so-called "low water automatic." This is a valve that cuts off the fuel if the water level in the boiler gets dangerously low and is normally positioned to operate long before the fusible plug becomes uncovered.

It consists of a series of expansion tubes arranged in parallel which, while all containing water, remain non-activated. But if the water level drops so low that steam enters one of them, this expands relative to the others and, in the process, closes a valve in the fuel line. They do not snap shut the fuel supply; but, for this purpose, this does not matter.

Unfortunately, the problem with this device is that without letting the water level get dangerously low, they are impossible to check for satisfactory functioning and, thus, must be considered only as an emergency back-up safety device which may or may not work!

Overhaul is fairly simple consisting of merely facing of the fuel valve seat and valve and checking the tubes for corrosion, plus attending to any glands that may need repacking.

Chapter 6
Superheaters

So far, we have considered steam being taken directly from one of the types of boilers described to the engine, where it produces mechanical movement. This steam is known as "saturated," as it inevitably contains some water particles from the mass of water from which the steam originated. The temperature of the steam so produced depends upon the pressure within the boiler. The higher the pressure, the higher the temperature for boiling point and, therefore, the outlet temperature of the steam.

We all know that water boils at 212 degrees Fahrenheit, but there should be a qualification to the effect that this is at atmospheric pressure at sea level. What is not so generally known is that if the pressure is increased to, say, 85 pounds per square inch on the gauge, the temperature of the steam produced at this pressure will then be 327 degrees F. Further increase in pressure results in a corresponding increase in temperature and at 250 pounds per square inch on the gauge, the steam produced will be at approximately 405 degrees F. This pressure is probably the highest you are likely to encounter in a shell-type boiler.

Pressures over 300 pounds per square inch are considered uneconomic on small, simple boilers, as anything higher needs such an increase in the cost of the material required to stand such pressures (not counting the additional maintenance problems that ensue) that the gain is not worth the cost.

Having now digested this rather alarming piece of theory—unless you habitually wear an asbestos suit—we will progress to the reasons for the superheater.

The superheater is a device which may be considered an extension to the heating surface of the boiler used to increase the temperature of the steam without increasing the pressure. During this process the volume is also increased by up to approximately 30 percent at 250 p.s.i.

This gives three main advantages: namely, conversion of the entrained water into steam; reduction of condensation in the engine; and a greater volume of usable steam without a corresponding increase in the demand on the boiler. The increase in volume is partly due to drops of water being turned into steam. Here, we add the second piece of alarming information—one unit volume of water when turned to steam becomes 1642 unit volumes.

Danger of Steam

The reason behind the warnings on safety when dealing with a steam plant, thus, become abundantly clear. Failure in a boiler containing any significant pressure immediately reduces that pressure to zero. The reduced pressure means that the water boils at a lower temperature and, thus, all water in the boiler turns immediately to steam, resulting in an almost instantaneous expansion by 1642 times in volume of the contents, causing the destructive explosion. Think about it—one bucket of water becomes 1642 buckets of steam, instantly!

Superheater Justification

Elimination of a few drops of water may not seem a great advantage, but when you consider that the speed of steam in a small pipe, say, under 3-inch diameter, can be 40 to 45 miles per hour, the reduction in wear to components no longer subject to the scouring action of water droplets can be considerable. It was worthy of consideration in a commercial undertaking where the plant was probably in continuous operation. As far as a preserved steam plant is concerned, the length of operation is likely to be relatively short and, thus, if a superheater either does not exist or needs a good deal of restoration, the work and cost involved are probably not justified.

There is one important exception to this. That is the case of transport steam vehicles, be they road or rail. When these are used, they are usually still required to produce their full original performance; at least for the limited time of operation.

Some firetube-type boilers have a superheater housed in a series of enlarged tubes which are known as "flues." The superheater itself consists of two small drums or "headers" connected by a series of small diameter (approximately 1 to 1-1/2-inch) tubes formed into elongated "U" shapes and projecting from the header inside the flue towards the furnace. Steam from the boiler, via the stop cock or regulator, enters the first or "wet" header from whence it has to pass through the superheater tubes or elements in order to reach the output or "hot header."

In passing through the elements, heat is absorbed from the gases in the flue, the resulting addition to the temperature of the steam being up to

Fig. 6-1. Stanley 10 hp engine exhibiting straps that anchor it to the rear axle.

300 degrees F. From the hot header the steam travels to the engine through large bore pipes, which should be as smoothly routed as possible to reduce frictional losses. (See Fig. 6-1.)

The elements, in particular, are subject to a tough life, being attacked both internally and externally by the respective gases.

The actual superheater, including headers, can be subjected to a static cold-water test much the same as applied to boilers. If replacement tubes are found to be necessary they should be of solid-drawn, mild steel in order to withstand the temperatures involved. These tubes are expanded into the headers in the same way as smoke and water tubes in the boiler.

The headers can be mild steel fabrications or castings and may be of circular or square cross section. The header is usually fitted with plugs to give access to the ends of the element tubes. The original design should be followed as this will provide a suitable outlet temperature. Do not be tempted to extend the elements too far towards the furnace as they will burn away very rapidly and the high temperatures produced before this happens will probably cause lubrication and/or distortion difficulties in the engine consuming the steam.

The same sort of configuration is applied to large stationary water tube boilers except, of course, that the elements do not live in flues, but in the space between the water tubes and the flue outlet. In flash boilers and the smaller type of firetube boilers generally fitted to steam cars, the superheater usually consists of one long tube coiled around in either the outlet space between the tubes or generating coil and the chimney, or between the burner and boiler.

Flash Boiler Superheaters

In the case of the flash tube or boiler, the superheater coil is, in fact, really just an extension of the steam generating coil. Some examples were actually all in one in the furnace, with the regulator intervening, of course. In these applications, the rest of the plant was normally designed at the outset to consume steam at high degrees of superheat.

As before, these superheaters must be made from solid-drawn steel tube or sometimes, particularly with the furnace type, an alloy steel. The original specifications should be tracked down, if at all possible, or the advice of a specialist organization sought, such as a steam engineer or one of the manufacturers of such tubing. In the latter case, tell them the use for which you require the pipe and the conditions appertaining thereto.

When renewing the coiled type of superheater you will have the problem of coiling it. Half-inch bore thick wall tube, probably about 3/4-inch outside diameter, does not coil up that easily. If you have a long length, find a substantial fixed pillar about the correct diameter. Anchor one end of the tube to this pillar, then walk around and around the "mulberry-bush" with the other end until a neat, tightly wound coil is achieved. However, as you probably need 20 feet of tube for the coil, plus at least another five feet to pull on when you get near the end of the 20 feet, you will have to start with rather a long length of tube.

Another method is to have only the actual length needed for the coil. Clamp one end in a substantial vise, and then heat the pipe starting at the clamped end. As it gets red-hot, slowly coil it neatly around, working the heat and the coiling along together. It will coil quite easily when hot, and since in many cases it will get red-hot in use, no damage will be done by this method.

If the superheater coil is between the burner and the boiler, arrange that connections to and from its ends are made outside the burner or boiler casings. Some early designs of firetube boilers, with their superheater in this position, connected it to the throttle or regulator valve by using one of the firetubes as a steam tube. This was carried out by driving taper

Fig. 6-2. Stanley direct-coupled engine—the spur gear on the axle and mating gear on the engine crankshaft can be seen. The front copper cover of the engine protects asbestos wrapped around the cylinders. Steam pipe can be seen lagged (rear right of picture), with check valve on the cylinder lubricator oil pipe.

connectors into both ends of the tube and connecting the regulator to the top and the superheater coil to the bottom. We do not recommend this system, as it is prone to give trouble. If it exists, we suggest you replace it with an external pipe which, if well lagged, will not lose any heat worth bothering about. (See Fig. 6-2.)

To prevent elements from overheating when steam was shut off, an anti-vacuum or "snifting" valve was often incorporated in the steam pipe, particularly on stationary plants, between the boiler outlet valve and the hot header. This was similar to a check valve, but on opening admitted air to the pipe via a tee. When working, steam pressure in the pipe kept the valve shut to atmosphere.

The observant may have noted that several references have been made to protecting superheaters and not trying to improve their efficiency by getting them closer to the fire, yet one example described as being used in the steam buggy is right on top of the fire without any protection or snifting valve. The reason for this is the different method of operation of the stationary engine versus the average steam buggy. Most stationary plants keep their fire running all the time, even if they are oil fired rather than solid-fuel fired. Thus, when steam is not being used, the fire still exists to overheat any superheater tube placed too near to it. On the other hand, in most cases when the buggy is not consuming steam, the main jets of the oil burner are shut off. Nevertheless, reasonably frequent replacement of the superheater coil on those designs with it on top of the burner is wise.

Chapter 7 Chimneys and Drafts

Although a chimney is usually associated with a tall, possibly brick, structure belching forth black smoke over the countryside , the term can be applied to any passage conveying hot gases from their sources to the atmosphere. However, as well as a wayout for the products of combustion, after all the heat required has been extracted in the boiler, the chimney can, and often does, perform another function—that is, the induction of airflow through the burner or fire. (See Fig. 7-1.)

This can be achieved in two ways, namely natural or induced draft. Natural draft is caused by the height of the chimney, hence the towering structures associated with mills, power stations and processing plants. The same effect is present to a lesser degree in the comparatively short efforts found on traction engines and similar small mobile power plants. The draft effect is caused by the difference in the weight of air within the chimney as compared to the weight of a similar column of atmospheric air outside the chimney.

The calculations to determine the height of one of these giant jobs which serve comparatively easy-steaming land boilers, is complex. It has to take into account such details as a loss of approximately 2 degrees F. in the temperature of the exhausted gas for every three feet of height. Suffice it to say that a chimney 100 feet high will produce an air pressure at the inlet of just over four pounds per square foot. Talk of such structures may be mind-boggling to the average steam engine enthusiast, but quite a number of such plants and their associated chimney edifices have been preserved on both sides of the Atlantic, sometimes by a consortium of enthusiasts with delusions of grandeur.

Having explained the function of such chimneys and given a brief idea of the salient measurements, it is not proposed to try to describe the

Fig. 7-1. The haystack boiler is unusual and not often encountered these days. This is really a cousin to the kettle on the fire, being almost exactly that! A partly drum, partly spherical vessel sits on a solid-fuel fire. The riveted construction is quite clear as are the numerous stay bars that the design needs to hold it together under pressure. The device poised on the brick chimney at the right is the air damper.

restoration of same, as it is a job for the expert and definitely not for the do-it-yourselfer.

Induced draft is caused by some device causing more "suck" from a chimney than would be gained purely by its height. In fixed plants, this can be a fan arranged to suck up the chimney. The other type of unnatural draft is the forced feed or "blower." This is a fan arranged to blow up the chimney and can give up to about a six inch water gauge reading, while induced draft by sucking can give up to a ten-inch one or more. Water gauge readings are actually a reading of vacuum, but can perhaps more readily be understood by equating one-inch of water gauge to 5.2 pounds per square foot of pressure. As mentioned earlier, the boilers to which these methods are applied are usually stationary installations where steaming requirements are fairly constant, space is not at a premium and there is, usually, a handy source of power to drive fans.

Road Vehicle Chimneys

For the higher steaming rates associated with vehicular steam plants, where weight and space occupied become important, some sort of steam blast is employed. This may be in some form of a continuous jet, fed with steam directly from the boiler, or by use of the exhaust steam from the engine (if this is not required to be used again) that is fed into a condenser.

Sometimes both methods are used—virgin steam jet when the vehicle is stationary and, thus, the engine is not using steam and does not have an exhaust, and the exhaust steam when running. In both cases, the supply pipe is constricted slightly to form a nozzle and the alignment of this with the chimney is most important if correct drafting is to be achieved. The jet or blast nozzle *must* be absolutely concentric with the chimney and, in the case of vertical chimneys, absolutely vertical itself so that the emerging jet of steam completely fills the chimney.

The height of the nozzle in relation to the chimney will depend upon the height of the chimney itself and the bore of the blast nozzle. A reasonable method of determining where the blast nozzle will come if, for example the original is missing, is to use a template tapered at one in six, inserted with the wide end sitting in the top of the chimney, the pointed end being down towards the smoke box. For a preliminary setting, the blast nozzle should be positioned where the template just sits in the bore. In the case of a locomotive-type boiler, this will probably be some distance below the bottom of the chimney where it is mounted on the smoke box.

If this is the case, a further template will be required—this time the taper being one in three. These templates may be cut from stout cardboard or old cartons. Insert this revised template with the pointed end in the blast nozzle, which is temporarily held at the previously determined position. If the wide part of the taper extends beyond the diameter of the chimney, the latter must be extended downward towards the blast nozzle until it meets the sides of the one in three taper, that is, until it meets the base of this taper at a point equal to the diameter of the chimney. Though difficult to describe, if you draw it out, it will become apparent!

In the case of a traction engine or other type having a tall chimney or stack (Figs. 1-10, 1-11 and 1-12.) The one in six taper will probably locate the blast nozzle some way up the actual existing chimney bore, in which case the procedure with the one in three template is not required.

Some chimneys are parallel and some have a slight taper out towards the top. There seems to be no noticeable difference in the steaming of identical boilers fitted with either type, so it depends upon what the original had, or which type proves more pleasing to your eye. The main part, or barrel, of the chimney is rolled from flat steel plate usually about 3/16-inch thick with a riveted seam. This will, of course, mean finding some bending rolls and probably will be a straightforward job for a sheet-metal works. Welding has been employed on the seam by some restorers instead of riveting, as it is much easier. If the original had a riveted seam, you can even cheat over this and stick on dummy rivet heads to retain visual authenticity!

The steel chimney barrel is very prone to heavy attack by corrosion due to the smoke passing through falling in temperature to the point where acids are condensed out on the internal surfaces.

The base of the chimney where fixed to the smoke box of circular form is usually a casting and will not require much more than a good cleaning and, possibly, replacement of the holding bolts.

The top end of this type of chimney can be finished off in a variety of ways (See Fig. 7-2.), ranging from a simple half-round steel beading riveted around the rim, to an elegantly flared, polished copper "cap." The latter can usually be rescued from the remains of the old chimney and re-used after carefully knocking out any dents, burnishing out scratches and bringing back to the state of which you can be proud!

Sizing Blast Nozzles

Reverting to blast nozzles, little theoretical data is available to determine accurately what size they should be. The size of the engine and

Fig. 7-2. The Goldsworthy-Gurney steam coach of 1827 shows the boiler uncovered. The chimney is imposing at the rear.

the steaming rate required from the boiler have a great influence on their size.

Thus having restored your chimney and boiler and roughly located the nozzle as previously described, start by trying a size of nozzle which has proved or is already proving successful in operation in a plant similar to that being restored. Bear in mind that it is much easier to slightly enlarge the hole in the blast nozzle than it is to reduce it. Therefore, start off a bit smaller than you think appropriate and experiment under actual steaming conditions, drilling out little by little until satisfaction is obtained.

It is worth also considering the position of the nozzle as the hole size is altered. It may be necessary to move the nozzle up or down relative to the chimney. Therefore, it may be worthwhile making the connection between the nozzle and the pipe feeding it as a screw thread, so that there is some variation in height easily obtainable without having to remake the pipe length or change pieces each time.

Modifying for Efficiency

The small private-carriage-type of steam buggy seldom has an imposing smoke stack. In fact, it is often difficult to see exactly where the chimney is. Most of these would work a lot better if they did have a useful upright imposing edifice, like the aforementioned traction engine. Instead, the usual system of almost nothing does little to provide any vacuum over the burner and, hence, is one of the reasons for the poor performance and frequent blow-back failings of many buggy burners.

In the early Stanley type, for example, the flue gases, after getting up through the boiler firetubes, had only a shallow space above the boiler in which to collect themselves and find an exit in the shape of a narrow slot to the rear of the boiler, passing along a small duct over the top of the water tank, and then descending vertically through another small duct behind the tank to emerge under the car. This is really a highly inefficient system!

However, if you do not wish to depart from the original design and erect a nice, tall chimney piece in cast iron, you will have to make the best of the manufacturer's layout. In this case, it is suggested that you modify it as much as you can, without spoiling the original appearance, to give more flue area.

In the first place, give as much smoke box space as you can above the boiler. About the only way to do this is by increasing the depth, diameter being governed by boiler size. If you have to remake the body for any reason—such as rotten wood—you can take the opportunity to add on an extra inch in height. It will not be discernible from the original to any but the real experts on that particular marque. If the original body is sound, it may be possible to gain this extra inch by blocking up the seat in those cases where it is immediately above the boiler.

For safety reasons, we suggest you make the smoke box from

substantial steel plate, not less than 1/8-inch thick, preferably 3/16-inch, with the top and sides which are to extend down around the boiler sides for at least half an inch, all welded together as one. Make a secure fixing for this top to the boiler. Then at least this cover or smoke box may give you some protection from the blast of a boiler explosion, if you do have such a calamity.

It is often possible to increase the size of the duct leading out from the smoke box and you may well be able to increase its effectiveness by making it of venturi shape. Try to make the exit from the smoke box as broad as possible where it leads into this duct. A suitable fish-tail ending to the final portion of the duct or chimney may provide a measure of extraction or "suck" from the surrounding air as the vehicle travels along.

Smoke Box Maintenance

Every so often your smoke box will need cleaning, thus, it must be capable of being removed from the boiler or be situated so that gaining access to it is reasonably easy.

In the case of locomotive-type boilers there is a hinged door for access to the interior and the soot therein should be removed at frequent intervals. In the case of boilers fired on solid fuel, there will be found a quantity of small cinders in the smoke box. If left to accumulate, these will cause problems due to corrosion and the burning of the plates.

When replacing a smoke box or closing up access doors, the joints involved must be perfectly airtight or the chimney will merely suck in extraneous air from these joints. This is the easier alternative to dragging the air all the way through the fire and tubes, which is what you want it to do. Fire clay is used widely to achieve airtightness, as it is cheap and easy to obtain and to use.

On the locomotive-type smoke box, it is reasonably easy to get the smoke box door tight; but, in the road buggy where all sorts of peculiar covers or smoke boxes exist above the boiler, you may need to devise a better sealing system for said cover.

Chapter 8
Three Types of Burners

A multitude of burner designs have been used, but they may be grouped into three basic types: solid fuel, gas, or liquid fuel. In the latter case, the fuel is usually vaporized so, in effect, becoming a gas.

The simplest of all these is the solid-fuel burner, or grate. (See Fig. 8-1.) This is arranged internally in the boiler and really needs no explanation. The Victorian domestic fire was an example of a type of solid-fuel burner grate. Should the grate be missing entirely from your particular boiler, a new one will have to be obtained or made. For the bars and all parts of the grate, cast iron is by far the most practical material and, depending on the size and shape of the grate, may be cast in one piece or halves or individual bars. What is most important to remember is to allow plenty of space between the bars—at least 50 percent.

Underneath there will usually be a box-like receptacle with doors. This is the ash pan and it is most important for the correct functioning of the grate that there are no air leaks, such as would occur if there were rusty holes therein. The only air required is that coming in via the doors or dampers. A damper is a flap hinged and provided with a lever to control its movement to allow more or less air in under the grate. Check that these and the main doors are not distorted and that all levers work freely and properly.

If your solid-fueled boiler fire belongs to a road vehicle, it is as well to warn you that you may not be too popular amongst the local population if when the ash pan is full, you empty it behind you on the roadway.

The gas or vaporized-liquid burners are more complex. Basically, most work on the Bunsen burner principle with air and the gas being drawn into the mixing tube, hopefully mixed in the correct proportions,

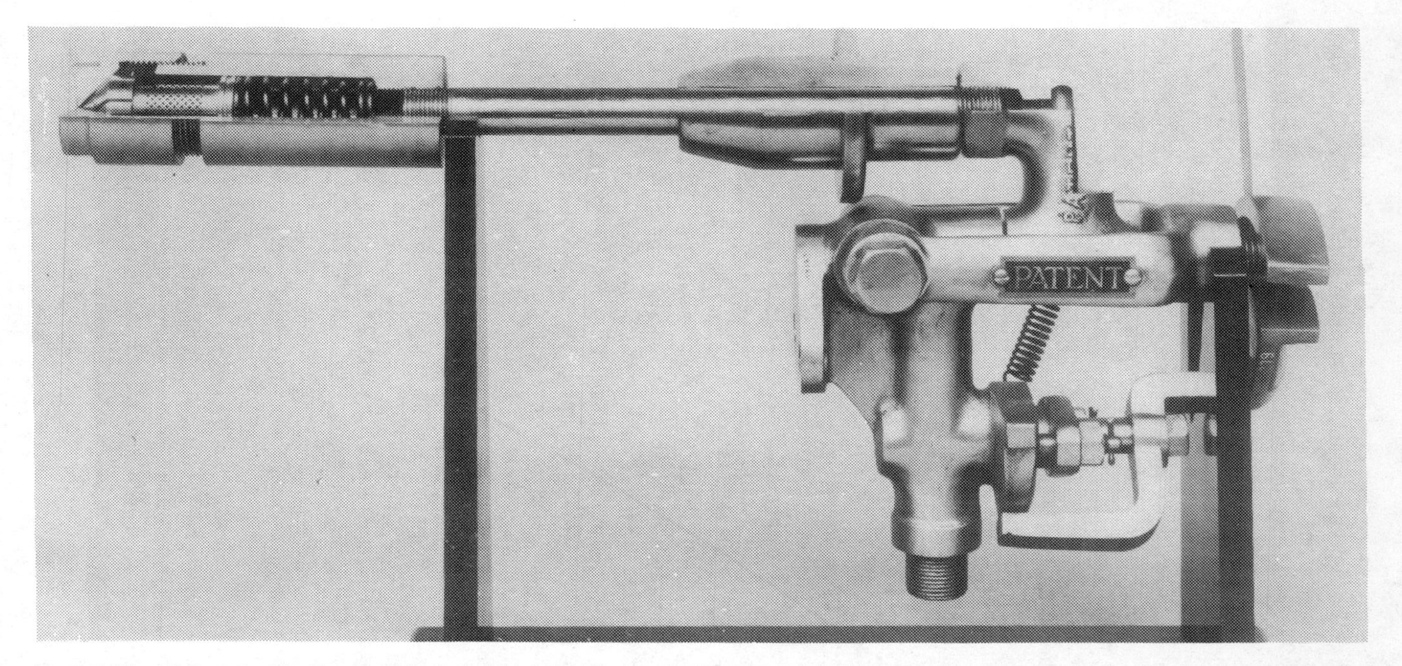

Fig. 8-1. The White patent fuel burner is typical of the type of burner that can be used with the heavier fuel oils, which works by atomzing the fuel and projecting it at some speed from the nozzle into the combustion space. Frequently, a special combustion chamber, designed to work with such a burner is used; but, it is also possible to use such burners to replace solid fuel in many ordinary grates.

and drawn upward through the burner and boiler by the chimney effect. The variations in design are legion, as are their problems. There seldom exists an owner of one of these burners with eyebrows!

The majority have a burner plate, often corrugated, round top or peaked corrugations, provided with a multitude of tiny holes or narrow slots, up through which a gas is persuaded to rise, usually by natural draft, but occasionally by forced induction. Herein lies the problem with most of this type—that is, those using liquid fuel, in that the liquid is loath to become gas or, when it does, it or its direct gas cousin declines to rise upward in the approved manner, but rather it blows back through the entry hole setting fire to everything in the vicinity, but failing to heat any water. This unfortunate habit is discussed further in the section on burner jets.

Overhauling Burner Plates

The first thing to do in overhauling a gas-type burner is to dismantle the entire assembly and extract the plate. This may be held in by screws, brackets, or simple gravity, but will almost certainly be packed with fire clay or asbestos, or both, which will have to be removed by whatever method you find most suitable.

With the plate out, carefully examine for cracks. If you find any, try to braze them up. Welding is not recommended as these plates are usually cast iron, a material for which there are few people sufficiently experienced in welding to tackle. Your average garage mechanic will undoubtedly say he can weld cast iron. So he probably can, after a fashion, but next time the plate gets hot, or even before, it is quite likely to crack on either side of the weld. The only successful cast iron welding the author has had carried out was on a cylinder block repaired by the Institute of Welding Research Laboratory. In this case, a special oven was built to house the item and the assembly was heated up very slowly over a period of seven days. It was then welded while still in the oven and gradually cooled for seven days.

After brazing up any cracks, any holes or slots blocked in this process must then be re-cut.

If your plate is either badly damaged or missing, you will have to get a new one. They are not that easy to find. Thus, you may have to have one cast. Provided you can borrow a good example, your local iron foundry should have no difficulty in casting, the most expensive part being the pattern although, fortunately in this case, this is not a very complicated design.

Your newly-cast plate will be rough around the edges and, thus, will need a few hours work with a file or rather less time with a power sander, smoothing it off. It will also need slots or holes cut.

Slots can be cut with a narrow or thin-bladed hacksaw, the only difficulty here being that you will probably be unable to get a blade long enough to go right across the plate and, thus, have to adopt peculiar angles and very short cuts to achieve the desired slots.

Holes are made with a drill. There is nothing technically difficult in this, but as you are having to drill into the sides of the corrugations at an angle, so that the holes point vertically upward, you will need to mount the plate on some form of jig. Even so, with the very small drills that are needed, you will undoubtedly break quite a few. The shape of the corrugation will govern the angle of drilling for the holes. The object is to make them sufficiently near to the vertical to stop the flame emanating from the side of one corrugation impinging on the flames from the side of the adjacent corrugation—and the best of luck to you! Other than a complicated drilling jig, hardly justified for one burner, there is no really easy way, other than sheer perseverance.

Under the cast plate one would expect to find a tin or copper pan whose function is to form the base of the mixing chamber, rather like the bottom half of a flying saucer, and which must form an airtight seal with the plate. Thus, after securing it, frequently by crimping around the edge of the burner plate, pack with fire clay. But, before doing so check that the distribution plates—discussed later—are, in fact, correctly positioned.

This pan unit then normally fits into an outer container, the sides of which are often duplex with the cavity thus formed filled with asbestos. This may be difficult to get nowadays because of the modern preoccupation with health hazards. However, try your friendly neighborhood plumber, who may well have some old asbestos cladding from central-heating pipes which can be broken up and packed into the casing. If possible, obtain a sheet of thin stainless steel for the inner ring of this casing—that is, the part nearest the fire.

Distribution Plates and Feed Pipes

Under the burner plate, and above its lower pan, usually will be found pieces of tin apparently covering a large portion of the holes or slots. These are deliberate; they are called distribution plates, and their function is to prevent the gas from coming straight out of the mixing tubes and up through the nearest holes. Thus, these plates must be so positioned that they are covering the ends of the mixing tubes. If this is not the case, it is often easier to extend the mixing tubes rather than alter the plates. The actual positions are not that critical, provided the mixing tubes do not extend more than two-thirds across the burner.

There are many variations of feed pipes to the jet or jets. Usually the fuel feed tube will enter the burner at the opposite side to the jets and coil around above the burner to act as a vaporizer. Such a tube or pipe may have one or more complete coils and may be duplicated for each jet. In all cases, make sure it is sound, not thin or corroded. High pressure fuel leaking straight on top of a large flame can provide far too effective thermal underwear. To vaporize properly, a fair area of tube is needed with walls not too thick, 13 or 14 gauge, and with pipe diameters between 1/4-inch and 3/8-inch O.D.—a range suitable for most burners.

Unfortunately, this gives too much fluid area and inhibits proper vaporization. Thus, to cut down the internal size of the pipe, a twisted steel cable—similar to that used for cycle cable brakes—ideally of stainless steel and nearly the full internal size of the pipe, should be pushed into each feed pipe for the full length of the portion inside the burner.

At the jet end try to warp as much as possible of the pipework outside the burner with asbestos tape, to retain the heat, and make sure all joints are really tight. The author prefers to silver solder all joints and possible sources of leaks, except for one disconnecting union. Any leakage encourages "lighting-back" outside the burner.

Jet sizes will be considered later, as setting up the burner fuel feed is often the most difficult part of any oil-fired steam machine.

Pressure Burner

Another type of liquid-fuel burner that may be encountered is that similar to a type still used today in domestic central-heating boilers. This is the pressure atomizing gun. In this design, fuel and, usually, air are fed in at high pressure and are mixed in the atomizer at the nozzle. The resultant gas is lighted as it emerges from the nozzle, which is inserted into a casing below the boiler, or in some designs with water tube boilers, in the center of the water tube/flash area. The advantage of this type is that there is virtually no risk of lighting-back and the correct mixture of fuel and air is supplied at all times. But, a means of pressurizing fuel and air is needed.

Also, when compared to solid-fuel firing, there is no messy ash or clinker to be removed and disposed of, or problems with having to introduce virgin solid fuel.

The oil may be fed by gravity from a high-mounted tank, and then pressurized at the burner. If a thick oil is being used, the tank may be heated by inserting steam-heating coils into it, thus making the oil less viscous and easier flowing.

Another method of feeding the fuel into the burner is to use steam from the boiler to inject it. A pressure on the order of 30 to 50 p.s.i. is usual. Consumption of steam power for this purpose may be from two to five percent of the boiler output. The rate of firing is regulated by a control valve in the oil side close to the burner, either manually or automatically operated. The air and steam pressure are often kept constant.

Due to the fine orifices encountered in the nozzle, the oil must be thoroughly filtered before use.

Chapter 9
Burner Jets: Mains and Pilots

The design of the burner jets is open to considerable variation and controversy—long nozzle, short nozzle, stubby taper, thin taper, with or without central cleaning wire or, even more sophisticated, an annulus jet allowing air through its center as well as round the outside. This latter type is more likely to be found, and more used, in the larger vehicle or stationary applications. Most buggy-type cars or small plants do not have a sufficiently large burner to warrant this design.

The jet is usually situated centrally relative to the air tubes of the burner. (See Fig. 9-1.) A useful venturi effect can be designed into the system which will greatly increase the flow of air into the burner—usually a commodity in short supply!

It is beyond the ability of the average amateur to calculate the relative quantities of air and fuel that are needed for, and can be drawn through, a particular burner and flue arrangement of the majority of liquid-fuel burners. Thus, we can only offer a guide of the range of sizes from which you can experiment when you attempt to raise steam!

Therefore, we will consider an average firetube boiler of 20-inch diameter by 18-inch height. A liquid-fuel burner for this could have either one or two jets. With a single jet, one would expect to have an air or mixing tube of at least 1-½-inch diameter; and with two jets, each tube would be around 1-inch diameter.

We would recommend starting with main jet orifices for a twin-tube design of size 63 or 64 drill diameter and for a single-tube design, about size 60. With such sizes, there should not be any difficulty in drawing in sufficient air, but it may be found that without a ridiculously high fuel pressure, it is not possible to get sufficient fuel into the burner to produce

Fig. 9-1. Front view of a circa 1910 Stanley steam car, with a so-called "Coffin-hood," showing the burner jets quite clearly.

enough heat to maintain steam pressure under load. Only an operational check will find this out.

Experiments to Gain Heat

If there is insufficient heat, first try increasing fuel pressure up to a maximum of about 120 p.s.i. If this does not increase the steam capacity sufficiently, try enlarging the jets one drill size at a time.

The problem is that the larger the jet, the more fuel is being supplied and, thus, more air is needed for combustion. Since the air-drawing capacity of the fuel jet is governed by the fuel speed, not its quantity, no more air will be drawn in for a given pressure, no matter how large the jet. In fact, due to the fuel occupying more area, a larger jet will tend to reduce the quantity of air.

Thus you will reach the point where there is insufficient air to burn any more fuel properly, and where a larger jet will either waste fuel, or more awkwardly, cause lighting-back below the burner plate.

With the largest jet that the air supply can accommodate, if there is still insufficient heat, you can either devise some form of forced draft or, where in the case of most antique plants this would be highly unoriginal, increase fuel pressure to increase the speed of the fuel and, thus, the amount of air drawn in. You may well need to reduce jet size down towards the original experimental size as, obviously, with increased pressure more fuel will be forced into the burner. High-pressure fuel demands great care with all joints and vessels under pressure. Two hundred pounds per square inch should be the absolute maximum. Keep well below this if you can.

It may be worthwhile experimenting with slightly different sizes of air or mixing tubes. But this usually means altering the burner and, thus, is not easy. Neither will it necessarily increase the air flow. In fact, it may prove counterproductive, as some of the venturi effect may be lost. "Big is better" is not always true.

What can be done easily and what sometimes helps is to give the air tubes a bell or taper mouth. Experiment with moving the jets in and out relative to these tubes.A starting point is with the jets away from the mouth of the tubes by an amount equal to the tube diameter. Generally, you will need to move in closer than this, but not always. Again, only trial-and-error can resolve this.

For a stationary plant, whether or not original, a forced-air feed is not difficult to arrange using electric fans of some sort; but, unless the design is relatively modern, a road vehicle is not likely to have any electric supply from which to run such a fan. As a steam engine stops moving when the vehicle stops, there is no permanently rotating drive that can be used for a mechanical hook-up.

In fact, a very few, relatively late, road vehicles did have forced draft, but the authors have not encountered any such beasts and can, thus, offer no advice thereon. No such mechanical devices have been noted on the early buggy-type steam cars. Their only form of forced draft is where the burner orifice points in the direction of travel—of marginal effect, in any case! (See Fig. 2-7.)

Pilot Jets

The pilot jet can nearly always be very small, between drill sizes 62 and 64 and, in some cases, even smaller. A strong pilot is essential to ensure the main fuel supply is vaporized at starting and when the main burner is shut down; but, there is no point in having a roaring pilot which will simply waste fuel. A pilot that will just about keep the boiler on the simmer when it is hot, is about right. For this very small jet, an air or mixing tube of about 1/2-inch diameter usually suffices. Some burners use only 3/8-inch with a relatively low pressure of about 20 p.s.i.

If your equipment uses a common fuel supply to both main and pilot jets, you may well find that originally both jets received fuel at the same pressure (probably around 50 p.s.i.) Thus, if you find you cannot get sufficient heat from the main jets at this pressure, the easiest scheme is to put in a separate tank for the pilot fuel. Feed the jet directly via a simple on-off valve, and dispense with any form of pump. Maintain pressure by using a tank capable of withstanding medium pressure. Fill this tank no more than two-thirds full. Pressurize by means of a tire-type valve, silver-soldered thereto, and a hand pump, to about 20 p.s.i.. Test this tank to at least double working pressure.

Alternatively, you can increase the pressure of your sole fuel supply and incorporate a pressure reducing valve in the take-off feed for the pilot. The author has a car with both systems, offering various permutations to be used in the event of failure of any one part.

If you have to make jets from scratch, a good basis for copying is the ordinary gas-welding torch jet. In fact, if you can get either undrilled or very small sizes it may be possible to use them as they are for jets.

Chapter 10
Fuels and Fuel Pumps

Solid fuel, as the name implies, is lumps of combustible matter burnt on a grate in a firebox. The most usual sort of this fuel nowadays is coal, although wood was, and sometimes still is, used where it is in plentiful supply.

For the best results, a steam coal or anthracite should be used, as this has the necessary heat output without producing a lot of ash and clinker. Coke was once a popular fuel, especially for road vehicles, due to its smokeless properties; but, with the decline in coal-gas manufacture, this fuel may not be readily available.

Apart from difficulties in acquisition, another disadvantage of solid fuels, particularly in road vehicles, is the space required for storage or bunkering. Also, in the case of coal, in particular, it is a messy substance calculated to do no good at all to your lily-white tuxedo.

Gas and Liquid Fuel

For convenience of handling and control, gas or liquid is preferable. With the huge increases in the use of butane and propane, there may be some advantage to be gained from using these gases; but, for adherence to originality the most likely type will be a liquid. The liquid used is usually a light oil that will vaporize easily, usually of the gasoline or kerosene range. Ordinary car fuel is not satisfactory unless it is of the *completely* lead-free type. Most car fuels contain a minute proportion of lead and, upon heating, this lead constituent precipitates out and deposits in the jets, blocking them. Even an entirely lead-free gasoline may not be satisfactory as, depending upon its constituents, it may not be sufficiently volatile to vaporize easily enough. At the end of this book, Appendix A gives data for a suitable fuel of this type.

The correct supply of liquid fuel to the burner is one of the most complicated of the steam car systems. This fuel must be maintained at a constant pressure, be perfectly clean, and be vaporized by the time it issues from the burner jet.

It is possible to pressurize the main fuel tank, but this is seldom done for several reasons: A very strong tank is needed if a high pressure is required, one has to depressurize to refuel and any leakage deposits rather a lot of fuel in quite a hurry in places where it is not meant to be.

Thus, fuel is usually supplied by a pump of some sort which may be a mechanical plunger-type (See Fig. 10-1.) driven from the crosshead, an electric rotary, or even a gear-type rotary mechanically connected to the driven shaft or axle.

Most burners require a steady fuel pressure. Thus, in the case of the crosshead-driven type, which of necessity is pulsating, it is usual to incorporate some pressure balancing device such as a small air pressure tank. As the amount of fuel needed depends on the vehicle's speed, load and gradient, the fuel pump must be capable of delivering more fuel than is needed most of the time.

To get rid of the excess, a pressure-relief valve is usually incorporated in the circuit. When excess fuel is being pumped this allows the excess to flow back into the supply tank.

Pump Rebuilding and Repairing

Nearly all types of pumps depend upon close tolerances and sound seals to work satisfactorily. In the plunger-type (See Fig. 10-2.) this means checking the fit of the plunger. A nice sliding fit is required—not too tight or it will heat up due to friction and bind. A slight seepage up the sides of the plunger lubricates it, and it will be kept in check by the seals or gland packing. For this, use P.T.F.E. or graphited string. If the plunger is too slack, these seals will not be able to contain the leakage.

If the pump body is worn oval, it should be reamed out true and an oversized plunger turned up from either stainless steel or bronze stock. Equally, if the pump body is sound but the plunger is scored, a new one must be made.

This type of pump needs inlet and outlet valves. Usually these are of the spring-loaded type, consisting of either a disc-, ball- or a mushroom-shaped valve, held onto its seat with a light spring. These valves are so arranged that the spring, in fact, does very little. Basically, the suction stroke pulls the outlet valve shut on its seat by suction, at the same time sucking open the inlet valve. On the delivery stroke, the reverse occurs.

For the valves, use brass or stainless steel discs, balls and mushrooms. If the valve seats are damaged, repair will depend upon their shape. Usually it will be necessary to make some form of end mill to chuck up in an electric drill, or turn by hand, to true up the damaged face.

Gear-type pumps, either mechanically or electrically driven, are usually of the two-spur gear-type, where one driven gear rotates a second

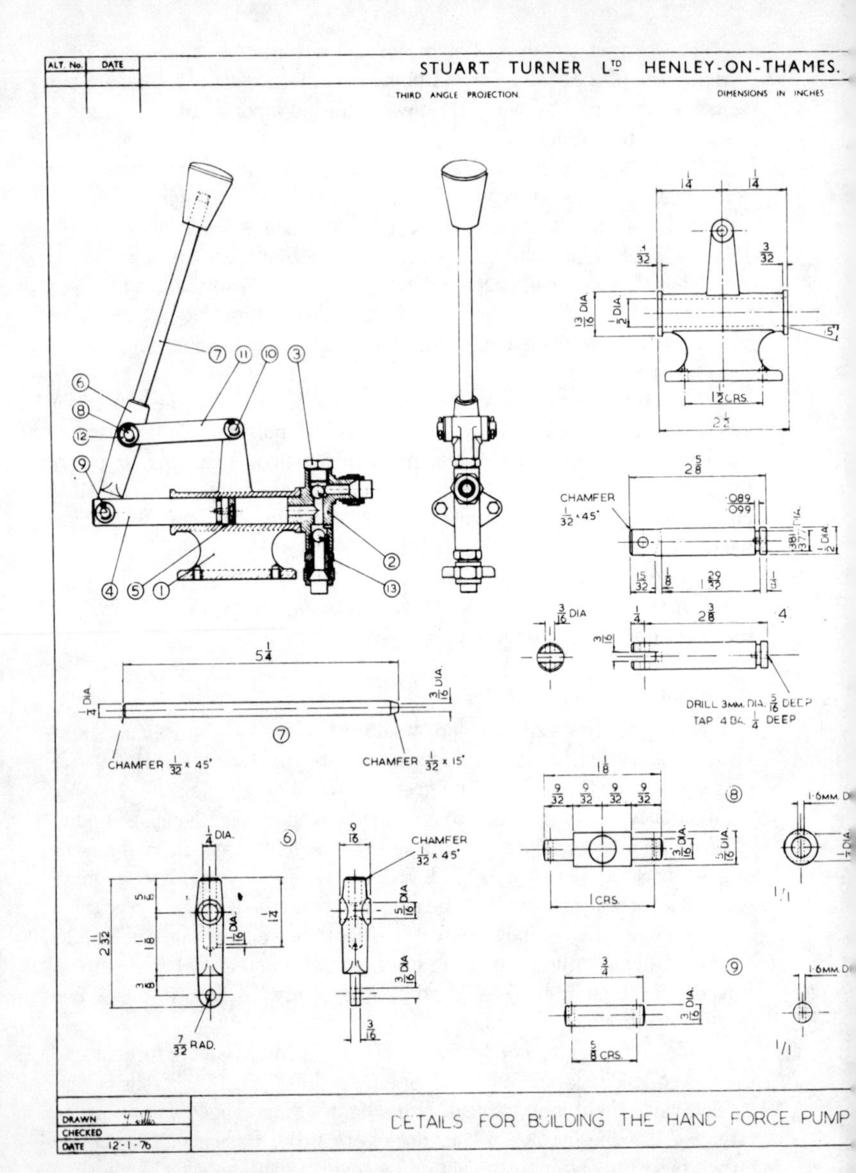

Fig. 10-1. Details of a typical plunger pump.

of similar size, the fuel being forced by their teeth out of an outlet port. In this design no valves are necessary. The usual fault found in these is excessive end play of the gears which allows fuel to escape underneath them and which can be rectified by facing the cover or the housing of the pump until a barely perceptible end float is present. An easy way to face

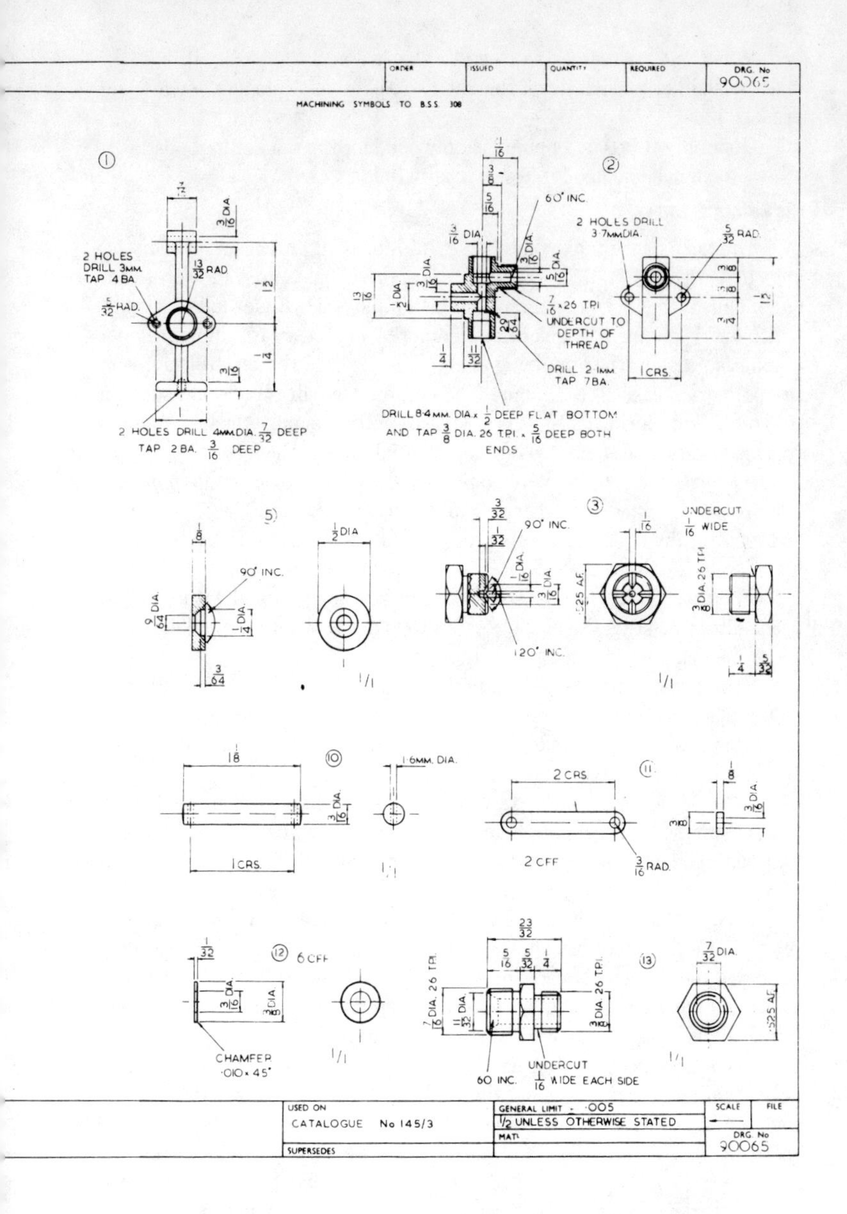

these parts is to rub them back and forth on a sheet of emery paper placed on a sheet of plate glass, thus ensuring a flat surface is achieved.

The driving shaft bearing must also be a good fit on the shaft. Any wear in either shaft or bearing means turning the shaft true and making up a new bearing from phosphor-bronze to be a good fit. Remember that the bearing needs a slight nip to retain it in the housing when pressed in; if you do not have a reamer to ream to size when fitted, you will have to make it a slightly slack fit on the shaft to allow for it closing down when pressed in.

Generally this type of pump is only used for the thicker fuel oils, as it allows too much leakage for low-viscosity fluids.

Air Balance Tank

The air balance tank can take several forms. In all cases, it must be able to withstand well in excess of the normal maximum fuel pressure used. One of the more common systems employs a cylinder full of air into which fuel is pumped at the bottom, compressing the air to half its original volume so that further pumping of fuel compresses the air still more on the delivery stroke of the pump. The compressed air retains pressure on the fuel feed during the return stroke of the pump, thus keeping a reasonably constant fuel pressure on the feed lines to the jets.

A variation on this system employs two cylinders, one containing just compressed air, and the other just fuel. The two cylinders are connected at their tops by a thin tube, air pressurizing the fuel in exactly the same way.

The idea of this was to stop air from becoming mixed with fuel and gradually exhausting the air pocket. In fact it is very difficult to keep only air in one cylinder and only fuel in the other.

Sometimes with the one-cylinder system there was a complicated baffle plate halfway up containing standpipes that trapped a layer of glycerine between fuel and air. In practices, the author has found this substance to be totally unnecessary.

All these cylinders are provided with some method of replenishing the air supply, and if this is pumped up once a day before starting up, it will run at least half, and usually a whole, day before the air becomes absorbed into the fuel.

A point that must be watched is that there must always be an excess pressure-relief valve that allows fuel to flow back to the tank if the pump becomes too enthusiastic. Otherwise it is possible to burst the balance tank and highly volatile fuel at, maybe, 150 p.s.i. or more is not the safest of materials. The author writes from experience. Having a tap which can shut off the excess pressure return line and one day having this accidentally shut, not noticing the build-up of fuel pressure, the balance tank burst, resulting in the rather wet seat of a passenger who happened to be sitting above it. The gentleman was not exactly pleased, especially since he presented such a fire hazard that he then had to walk home. Due to the various extra valves this particular vehicle incorporates, by manipulation of these, it was possible to isolate the burst tank and get home on a pulsating fuel supply.

Pressure Relief Valve

The excess pressure relief valve is, in effect, a non-return valve, normally provided with an adjustment for altering the pressure at which it

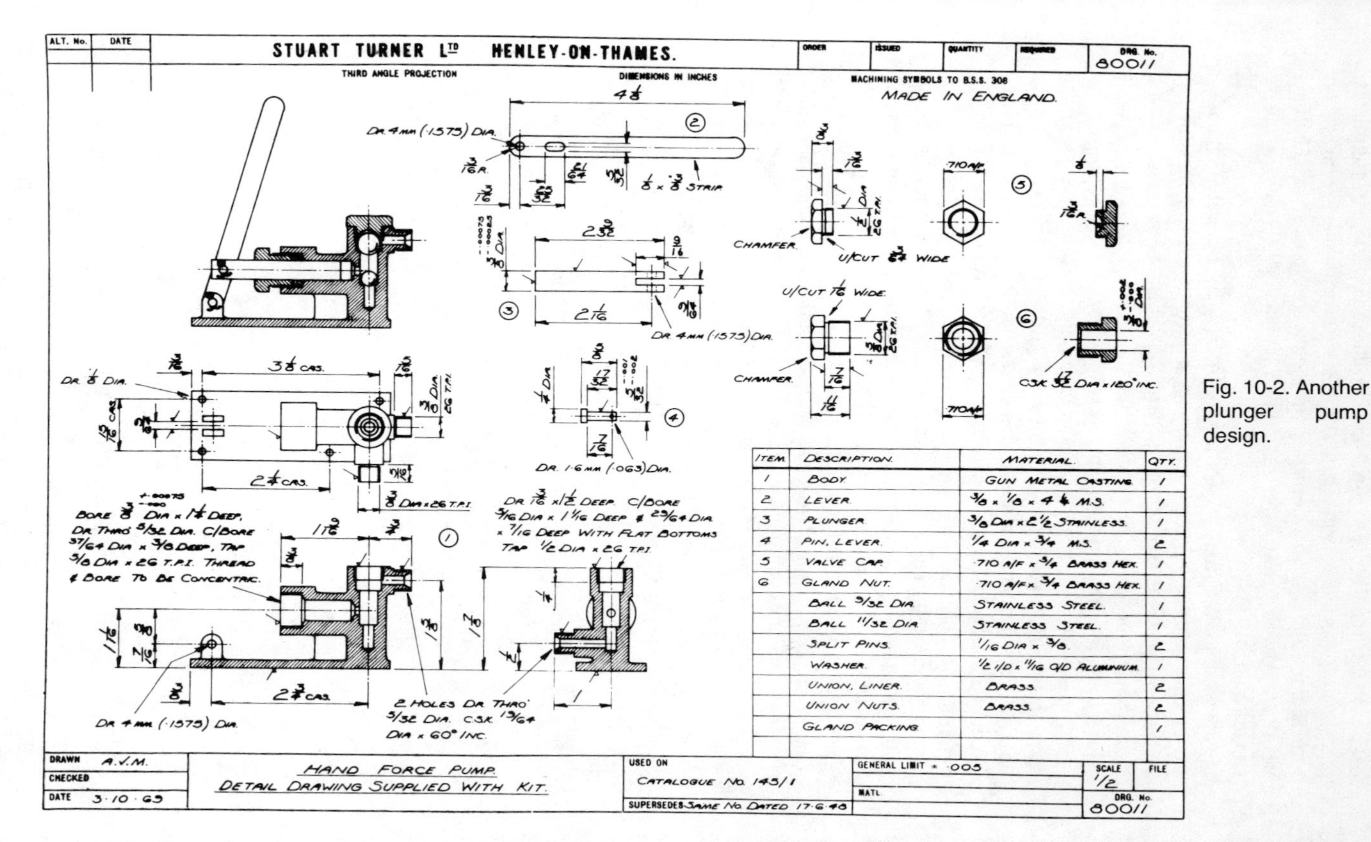

ITEM	DESCRIPTION	MATERIAL	QTY.
1	BODY	GUN METAL CASTING	1
2	LEVER	3/8 x 1/8 x 4 1/4 M.S.	1
3	PLUNGER	3/8 DIA x 2 1/2 STAINLESS	1
4	PIN, LEVER	1/4 DIA x 3/4 M.S.	2
5	VALVE CAP	·710 A/F x 3/4 BRASS HEX.	1
6	GLAND NUT	·710 A/F x 3/4 BRASS HEX.	1
	BALL 5/32 DIA	STAINLESS STEEL	1
	BALL 11/32 DIA	STAINLESS STEEL	1
	SPLIT PINS	1/16 DIA x 3/8	2
	WASHER	1/2 I/D x 11/16 O/D ALUMINIUM	1
	UNION, LINER	BRASS	2
	UNION NUTS	BRASS	2
	GLAND PACKING		1

Fig. 10-2. Another plunger pump design.

opens. It can thus be set to the desired maximum fuel pressure. The actual valve can be a ball bearing, a flat disc or a mushroom valve. Overhaul is similar to that described for the fuel pump valves and will not be repeated. However, the method of opening the valve is more complicated.

Sometimes the fuel pressure acts directly on the valve against a spring, but more often the fuel impinges on a diaphragm to which is connected the valve. This gives a more controlled opening and, thus, a more constant pressure. The snag is that this diaphragm is, in essence, a weak device which is prone to rupture.

If there is any sign of weakness or fatigue, make a new diaphragm from shim stock copper, well annealed, and if the design lends itself, add an extra return tube from above this diaphragm to the fuel tank. In the event of any leakage through the diaphragm for any reason, such fluid will return to the tank.

Since for different fuels you may well wish to adjust the maximum fuel pressure for the most satisfactory burner output, make sure this valve is located in an accessible position.

Hand fuel pumps are nearly always of the reciprocating-plunger type. Unless you like pumping like a maniac, make sure the diameter and, thus, the capacity of the pump are much larger than the power pump—at least two, preferably three, times the capacity. You can achieve the same result with a long stroke, but generally this is a nuisance, as by the time the leverage of a pump handle long enough is added, you have a ridiculous journey for your hand.

Chapter 11
Water Feed and Injectors

There are two main systems of feeding water to the boiler: first by pump usually reciprocal, and the second by injector.

If the pump system is used, it must be capable of pumping in excess of boiler pressure. Elementary, one might think, but this is a point often overlooked by many. This virtually rules out anything other than the piston-type of pump, which is usually of similar design to that type of pump described under the section on fuel pumps. (See Fig. 11-1.) That is, a cast body, often bronze, in which slides a reasonably tight-fitting plunger, again often brass or bronze, with a packing gland at its outer end to contain seepage past the plunger. Inlet and outlet ports both have non-return valves to allow the plunger to suck liquid into the body on its outward stroke and push it back out through the outlet port on the return stroke. These non-return valves are usually either ball bearings or taper seat or mushroom valves.

Sometimes an automatic by-pass may be employed, similar to a fuel system automatic, so that when the steam is at sufficiently high pressure the water returns to the tank rather than being pumped into the boiler. However, the author does not like this system and strongly discourages its use. It is quite possible for the steam pressure to be quite high when the water level is disasterously low in the boiler. There are other conditions of operation where it is necessary to add more water to the boiler, contrary to the normal requirements.

The preferred way is to have a manual by-pass valve which is opened when the water level in the boiler, seen via the sight-gauge glass, is sufficiently high. Normally, water is continually pumped into the boiler by the mechanical pump, and the by-pass valve only opens when more water

Fig. 11-1. Control gear in a 1904 Stanley Buggy, situated below the floorboards between front and rear seats: Front Compartment—(from left) cylinder oil lubricator tank, mechanical fuel pump, reverse pedal, hand fuel pump, (above it), main fuel pressure gauge and adjacent pilot fuel pressure gauge.
Middle Compartment—socket for hand water pump, changeover valve, whistle lever, hand fuel pump socket and handbrake.
Rear Compartment—hand water pump and rocking shaft for mechanical pumps.

is being fed in than is being converted to steam, or when it is required to increase steam capacity for a short while. The only problem with manual operation of the water supply is that it does need careful monitoring of the level gauge, as both too little and too much water can be disastrous.

It is not sufficient to rely on the pump non-return valves to prevent reverse feed from the boiler; a check valve must be incorporated as close to the water entry point of the boiler as is practical. This may or may not incorporate a shutoff cock. The advantage of having such a valve is that a leaking check valve can be dealt with while the boiler is under pressure. Check valves of various types are described in more detail in Chapter 14.

The usual layout is to run from the water tank via a stop-cock to a "T" junction. Branches supply both manual and mechanical pumps, which then converge their outlets to a common feed via the check valve to the boiler. This connection runs into a standpipe internally which disgorges the water near the base of the boiler. Between the pump outlets and the check valve, a "T" junction is usually provided to enable a by-pass valve to be connected in the circuit. When open this lets the feed water run straight back to the tank, such return being by means of a top entry in the tank.

Make sure that all connections, particularly those on the outlet side of the pumps and at the boiler, are absolutely first class. There is no room for doubt in these. *Any nipples on pipes should be silver-soldered, not soft-soldered.*

Injectors

Probably the simplest device for feeding water into a boiler is the injector. Its very simplicity however has caused it to be the subject of myth and mystery, as from superficial examination it is not obvious how the device works. Before tackling the overhaul of these most useful implements, it is, therefore, proposed to explain just how they work.

Reference to Fig. 11-2, will show that the body of the injector contains a series of converging and diverging cones. From the right are the steam cone, combining cone, and delivery cone. Projecting into the steam cone is a tapered spindle capable of being moved in or out by a screw thread and operated from the small hand wheel on the extreme right of the picture. This serves to vary the orifice of the steam cone and, therefore, the amount of steam able to pass through. This arrangement is not often found on modern injectors, but was encountered from time to time in earlier days and could well be found by the restorer. Careful sizing of the cones, in relation to each other and to the working pressure performing the function, is required.

As a matter of historical interest, it is believed that Sharp, Stewart & Co., the famous locomotive builder, obtained the concession to manufacture injectors to Gifford's patent, due to the fact that they were the only firm to make a working example from his drawings without being told how it worked. So, how does it work?

The steam is admitted to the steam cone via a stop valve mounted on the boiler, towards its top. As mentioned before, the size of this cone controls the amount of steam used. There are two more functions to fulfill, however—one is that the emerging jet of steam is directed in precisely the right direction, and the other is the degree of converging taper. This is the ingenious bit, as the pressure energy of the steam is partly used to increase its velocity. That is, the jet of steam emerges from the steam cone at a higher speed, but at lower pressure, than it had on entry. Around the entry to the combining cone is the water admitted from the feed tank. This enters through a comparatively large bore and is, therefore, moving slowly.

In the combining cone (convergent like the steam cone) the slowly moving cold water combines with the swiftly moving jet of steam which is condensed in the process. It is worth mentioning at this point that the colder the feed water, the better the condensation of the steam. The combining cone should ensure that all the steam is actually condensed and combined with the feed water prior to delivery to the boiler in the form of a solid jet of hot water traveling at high velocity. Again, the cone enables energy to be extracted from the steam and given up to accelerating the hot water.

The next process has to reverse this trend and convert velocity energy into sufficient pressure energy to force the water into the boiler against the steam pressure contained therein. This is done by the delivery

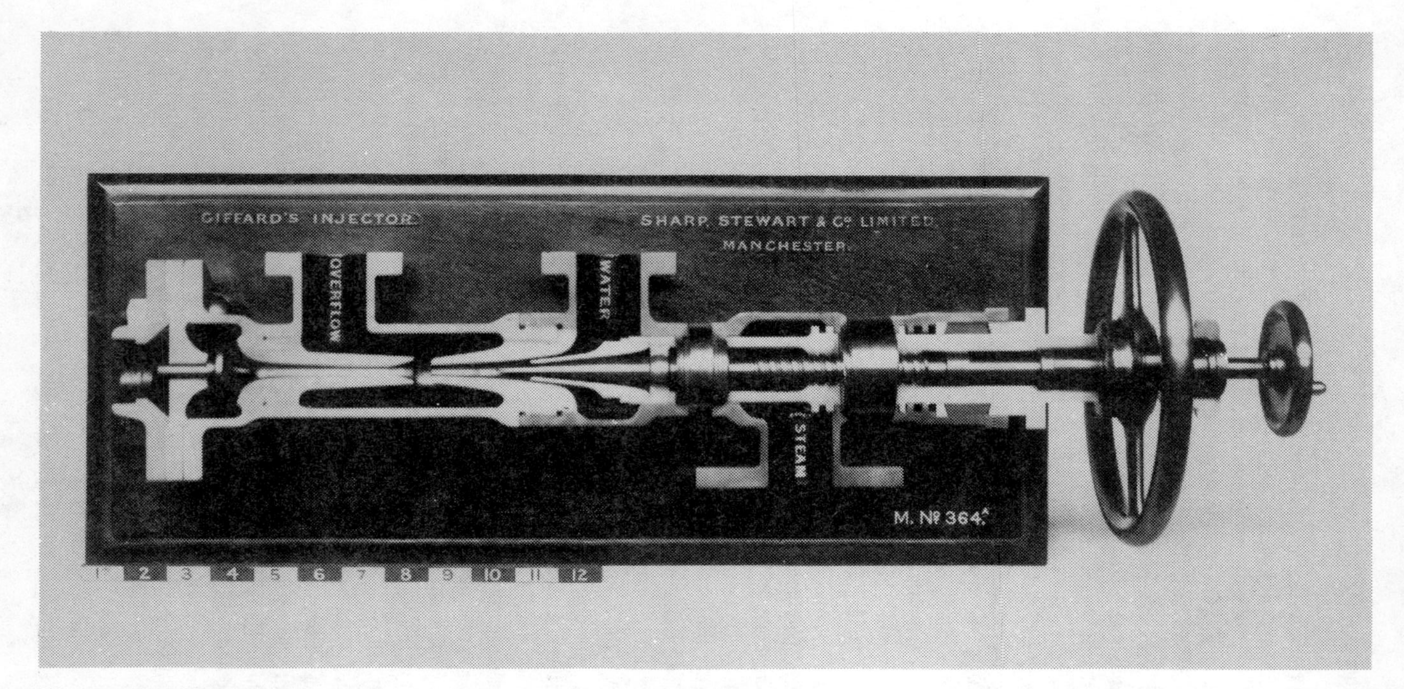

Fig. 11-2. The Gifford injector, in section, is of the adjustable type, the various components being clearly shown. From the right: steam admission adjusting wheel, the steam cone, the combining cone, and the delivery cone.

cone of divergent form, the flow of hot water gradually being reduced in velocity, but increased in pressure.

The more observant will have noticed that from the gap between the combining cone and the delivery cone there leads an outlet marked "overflow." This should only be used during the starting of the injector until the flows become established. The overflow will also spout water if there is any interruption to the flows from the cones while the injector is working. Continuous discharge from the overflow indicates the need for attention, provided always that the pressure, supply and other conditions are normal.

Figure 11-3 illustrates a more modern type of injector in which it will be noticed that the combining cone is in two pieces, the left-hand part being free to slide away from the right-hand part. This gives the injector the property of being automatically restarting if, for any reason such as vibration, the jet is interrupted and the injector stops feeding. The cone than opens the overflow while excess water is expelled. As soon as the vacuum reforms, the cone slides together and becomes one again.

Overhauling Injectors

There is not a great deal one can do to overhaul these devices. Basically, one can dismantle, clean and reassemble them. The easiest way to describe the procedures open to the amateur is by a series of notes. If these do not work, a new injector or a professional overhaul is the only answer.

1. To dismantle the Gresham & Carven-type injector, there are only four detachable parts. Referring to the figure, the left-hand nozzle plug and the left-hand cone assembly can both be removed by undoing their respective hexagons. The right-hand hexagon undoes the steam cone assembly, leaving only the combining cone which is held by a setscrew. After removal of this, the cone can be pushed out to the left.

2. The Giffard-type is slightly different. The threaded spindle and various gland nuts can be undone and removed, followed by the left-hand cover plate. The delivery cone comes out by unscrewing the left-hand end of the injector body, whereas access to both combining and steam cones is by unscrewing the center of the body. Try not to damage the body unions in the process.

3. Keep the cones clean. They will fur up rather like a domestic kettle, especially in hard water areas. Use household kettle de-furer to clean. *Do not poke up the cones with bits of old iron or any other material to clean them,* as all that you are likely to do is to enlarge the throat size and, thus, lose efficiency. It is surprising how much even a piece of wire can damage a carefully machined venturi.

4. When reassembling, it is most important to ensure there are no burrs or specks of dirt that can throw the cones off center. Satisfactory operation of the injector depends entirely on all cones being exactly in line as made. It is a good idea to use a little liquid jointing on the threads, such

as Boss White, but only on the threads. None must be allowed to get on the smooth, sliding fit surfaces or, of course, on the actual cones.

5. Keep all connections tight, especially the water inlet, as any air admitted here will prevent the device from working.

6. The feed water should be kept as cool as possible. The hotter the water, the less efficient the injector. On the same theme, a slightly leaking steam shut-off valve will allow the injector to become too hot to start. A simple cure here is a bucket of cold water thrown over the thing.

7. Ensure that the valves supplying steam and water to the injector open fully and do not restrict the supply in any way. Also, try to avoid sharp bends in the pipework connections to and from the injector.

8. If a filter is fitted in the water supply, ensure that it does not restrict the supply in any way. A filter is recommended to prevent the ingress of sundry impurities and items such as small frogs and fish if the water is picked up "in the country."

9. Injectors are classified for size by throat diameter at the smallest part of the delivery cone. As far as the authors know, every country in the world expresses this size in millimeters. Catalogues usually quote a feed rate per minute for injectors. Thus, if you do not have the original, you will have to try to decide what feed capacity you need before looking for an injector. Better still, quote your boiler capacity to your favorite injector maker and ask him to tell you what size injector you need.

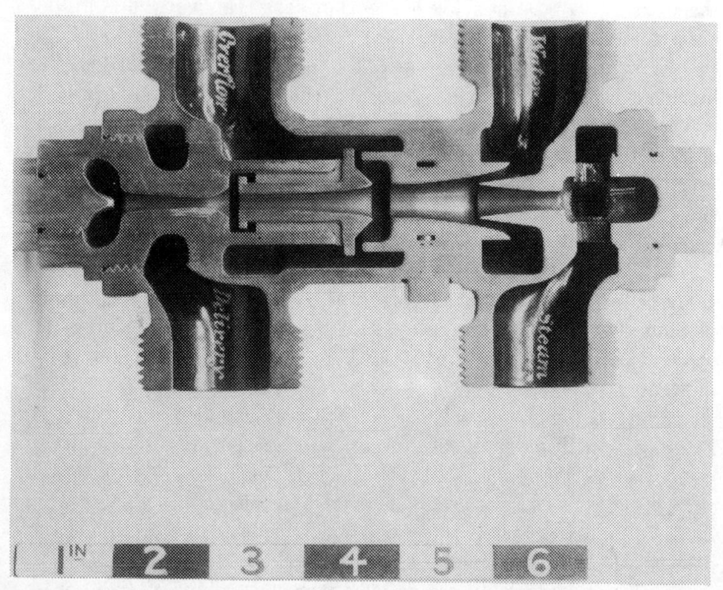

Fig. 11-3. Gresham & Craven restarting injector of about 1885, showing from the right: the steam cone, the combining cone, and finally, the delivery cone. The construction and method of dismantling are quite clear from the sectional view.

Another use sometimes made of injectors in road vehicles is the use of one to fill water tanks in conjunction with a long hose into a convenient stream or horse trough. If possible, keep the supply water tank above the height of the injectors. Although injectors will suck up water, their life is much easier if the water can flow down into them by gravity.

Feed Water

Finally in this chapter it should be drawn to the reader's attention that just any old water will not do. Particular attention must be paid to the quality of the feed water fed to the boiler or all the good work previously put into the restoration will have been in vain.

There are various impurities which, if present in the feed water in any significant quantity, will be detrimental to the boiler. The most common of these are lime (referred to as "hardness" and discussed at some length in Chapter 20), air, salt, and sediments of earth, mud or clay.

Air is found in all natural water and has advantages and disadvantages. The air present in the water slightly lowers the boiling point of the water and tends to cause a more gradual generation of steam; but water containing air also has carbonic acid present, which causes corrosion.

Salt is associated chiefly with sea water, where its proportion averages approximately three percent. Salt and other solid impurities in the feed water are very harmful to the boiler as they are deposited on the tubes and plates, leading to a build-up of scale. Such deposits retard the heat transmission from the gases to the water, resulting in poor steam generation and possible burning and weakening of the tube or plate. A ⅛-inch deposit of scale has been calculated to impair the effective work of the boiler by up to 25 percent. The moral here is to avoid using sea water as boiler feed—a lesson learned the hard way by early steamship operators!

Other impurities present in the water will depend upon the source of that liquid, the best being that collected by surface drainage into a reservoir, as solids will have tended to settle out. River water can be variable in quality depending largely upon whether any industry is present upstream of the collecting area. Various acids may be present, but these can be encountered by the addition of proprietary feed-water treatments. (See Appendix A.) Acids, of course, attack the metals in the boiler resulting in pitting and corrosion. Regular blowing down of the boiler will help to prevent impurities from building up too high a level.

Incidentally, the use of pure water does not solve the problems associated with the containment of air, as it boils at a higher temperature and steam generation takes place with great violence. This causes a solvent action on steel and can lead to serious corrosion and, therefore, weakening of the plates, thus, shortening the effective life of these parts with possibly disastrous results.

Chapter 12 Tanks: Water, Fuel, Air and Oil

We are not talking about things with steel tracks, where there is a great deal we could talk about, but rather hollow things that hold liquids, about which there is not an awful lot to say.

There are several different uses for tanks in steam plants. Nearly all have water and oil tanks. Those with liquid-fired burners also need a fuel tank and, in many designs, an air and/or pressure balance tank.

Water Tanks

Water tanks are best made from copper; but for economy and, particularly, for large, external stationary plants, galvanized steel is often used. Wherever practical it will help to save fuel and produce steam more easily if you can position the water tank close to or even around the boiler, thus giving it a measure of preheating. Some of the early steam buggies had a horseshoe-shaped water tank around the boiler. (See Fig. 2-10.)

Fittings required in a water tank include a filler orifice of reasonable size—not less than 2-½-inch diameter and preferably bigger. You do not know what you may have to use to fill it in an emergency. You need a water level or contents gauge, usually mounted vertically adjacent to one side. Underneath, you need a drain plug plus a supply outlet with stop cock, which may or may not incorporate a filter.

Fuel Tanks

The fuel tank can be galvanized steel or copper. Oddly enough, galvanized steel (not leaded steel) is to be preferred as the petro-carbon fuels used tend to react with a copper tank to form, over a long period, a gummy substance that blocks up fuel lines and eventually jets. This tank needs a filling orifice big enough to take a funnel easily, a drain plug, and a supply or feed outlet (which sometimes is one unit with the drain plug, in which case it usually has a three-way cock with center position "off"). Most designs also need a union in the top of the tank to take the return of excess fuel from the fuel-relief valve.

If the supply union does not incorporate a filter, one must be added somewhere in the fuel line. Try to use one of the correct period for the plant. Although the principle is the same as for modern gasoline car devices, these look wrong. If the original is missing, search for one of all brass construction, usually with a brass bowl rather than glass. Inside may be a series of discs held by a nut on a central stud. Take them all off, clean each carefully and replace. It may have a wire gauze filter which will probably be damaged, so make a new one. If it is gummed up, even if undamaged, it will probably be easier to make another than to clean it.

Pressure balance tanks are discussed in the fuel section, but are usually cylindrical tanks with domed ends, capable of withstanding quite high pressures of up to 200 p.s.i. They are normally of quite small capacity, no more than two pints and often much less. (See also Fig. 14-1.)

Air and Oil Tanks

Air tanks are sometimes needed to pressurize various parts of the fuel feed system, and these must be made to the same standards as industrial compressor tanks. Usually these only need a pressurizing valve, an outlet union with stop cock, and provision for connecting an easily seen pressure gauge. (See Fig. 12-1.)

Oil tanks can be any convenient shape, are nearly always arranged for gravity feed to the pump unit, and need to hold not less than half a gallon of oil. Since cylinder oil is thick and sticky stuff, they need a reasonably sized filler orifice, not less than 1-inch diameter. Apart from this, they need only an outlet union with a stop cock. (See Fig. 11-1.)

The size and shape of each of these tanks depends upon the space available for their location. For a non-condensing steam car you will probably use around one gallon of water per mile, thus anything less than 20 gallons is pretty useless. If you have to make a new tank, it is well worthwhile considering the possibility of finding enough space to make it big enough to hold 30 of 40 gallons. Often this can be done by filling irregular spare spaces or perhaps increasing overall depth, if there is room. Fuel consumption can go as high as five miles per gallon, and you will be lucky if it goes below 10 m.p.g. Thus, if making a new tank, allow for at least eight gallons. Remember, it is much easier to find water than fuel on your route!

Tank Repairs

Maintenance or overhaul of existing tanks usually consists of repairing any corrosion holes. These are hardly ever found in oil tanks, but are

Fig. 12-1. Separate fuel tank for pilot fuel, which may be either gas or liquid. Top left of tank is the air pressure valve, air being supplied by either a hand or foot pump, or a garage supply. Below the tank is the main stop cock, used when the vehicle is not in use. Also shown is the pressure gauge recording about 20 p.s.i., the usual operating pressure.

found frequently in water and fuel versions. We would remind those who may be surprised at fuel tanks rusting out, that even if the fuel does not contain any water (which it, in fact, often does), the atmosphere most certainly does, unless you live in Nevada. This damp vapor condenses in the tank and, being heavier than the fuel, sinks to the bottom where, in due course, it finds some air and produces rust.

When repairing holes in fuel tanks, do not apply a naked light to an empty tank. Even if it has been empty for some time, the chances are that there will still be some fuel vapor present which will be only too willing, given the chance, to demonstrate a miniature version of a boiler explosion. Fill the tank with water, obviously upside-down if the holes are in the bottom; then braze, weld or solder, whichever you prefer, while it is full. If the metal is badly corroded, it will probably be easier to make a completely new bottom. After cutting out the rotten metal, stand well back and apply a flame to the interior to burn off any remaining vapor. You may find, if corrosion is extensive, it is easier to make an entirely new tank.

Copper water tanks do not usually give so much trouble. You may get a certain amount of corrosion, but you will be unlucky if this is sufficiently serious to warrant replacement of a panel. However, copper fuel tanks are

Fig. 12-2. A typical turn-of-the-century "Buggy" style steam car body was wood frame, paneled either in wood, wood and metal, or all metal, depending on the make and had no separate chassis.

a different matter. Although they do not corrode, as mentioned previously they do tend after a time to suffer with the gum problem. This is extremely difficult to remove. The authors would be most interested to know of any chemical that has proved successful in removing this gum without damaging the copper, particularly in tanks that have been idle for several decades. Unless such a substance is available, about the only alternative is to cut open the top surface with a large enough hole to admit a hand and a wire brush and then apply hard work.

When fixing tanks, remember that two dissimilar metals when touching each other react electrolytically to cause corrosion. If you are using steel straps or brackets to hold down a copper tan, interpose a neutral substance between them, such as a piece of felt.

Water and fuel tanks situated in wood-bodied steam buggies (See Fig. 12-2.) often do not need to be fixed down, as the closeness of their fit in the space allocated to them is such that they cannot move. However, in stationary plants this does not usually apply, and they must be fixed.

When making new tanks without the original to copy, one small but important point should be born in mind: Make sure you put the filler where you can get at it to fill with the required liquid!

Chapter 13

Condensers = Conservation

The steam exhausted from the engine is usually released straight into the atmosphere via either a pipe or a chimney and blast nozzle arrangement. In certain applications, however, it is advantageous to retain this exhausted steam for further use by condensing it back into water. (See Fig. 2-10.)

Why Condense?

In the case of a vehicle, it is obvious that the amount of feed water that can be carried is limited. This restricts the range of the vehicle unless some of the exhaust steam can be used again—that is, condensed, to increase the range between water replenishment. The most usual sort of condenser for this application is similar in appearance to the radiator used with internal- combustion-engined vehicles.

Small header and lower tanks are connected by numerous finned tubes between which air flows, either naturally or with the assistance of a power-driven fan. The fins on the tubes are used to increase the surface area exposed to cooling air. In action the exhaust steam is led to the header tank from whence it goes down the tubes, being cooled in the process, hopefully, back to water in the bottom tank. From here the now-hot water is conveyed to a hot feed tank from which it is pumped into the boiler. As well as reducing the necessity to stop so frequently to fill the water tanks, this feed has the advantage of being hot when entering the boiler, thus, requiring less fuel to convert it to steam than is the case when using virgin cold water.

This condenser is not subject to great pressure; but for successful operation, there should not be any leaks in the system. Soldered construction is usually employed. Thus, you can either repair it yourself, if you are reasonably adept with a large soldering iron or small blow lamp, or persuade your friendly gasoline-engine car radiator specialist to deal with it for you.

The quest for economy of operation made condensing an essential addition to the very large plants installed in mills and similar large power-consuming industrial concerns. It was not a question of any water supply problem. Usually there was adequate water and space to store it, but there was the significant saving of fuel by using heated feed water, plus the increase in power that was made available by the use of a condenser operating under a vacuum.

As the volume of steam in a vessel condenses, the small quantity of water thus produced from a much larger volume in area of steam leaves a void or partial vacuum. In the case of stationary plants this vacuum is made as high as possible by means of an air pump driven by the engine. Vacua on the order of 26 to 28 inches of mercury are usual. As one inch of mercury vacuum is approximately equal to half a pound per square inch of pressure, an engine operating at, say, 28 inches vacuum will have, in effect, an extra 14 p.s.i., added to its working pressure, plus the loss of back pressure which can be six to eight p.s.i..

The type of condenser used is of horizontal form with a chamber at each end connected by tubes similar to boiler tubes. They are expanded into the tube plates in the usual way. The end compartments have access doors while the portion between the end tanks, the tube area, is enclosed in a chamber which is characteristically pear-shaped (inverted). Cold water is circulated through this chamber, cooling the tubes and condensing the steam.

This condensed water is extracted by a condenser-pump and used for boiler feed as before. The condenser is constructed of iron castings and steel plate, with tube corrosion being the biggest problem. However, as no great pressures are involved, straightforward fabrication can be employed to renew any damaged or scrap parts. Sometimes good sections of withdrawn boiler tubes can be employed as replacement condenser tubes. Again, in order to maintain vacuum, all the components and connections must be airtight.

Oil In Condenser

One important consideration with a condensing plant using the condensate as boiler feed water is that the exhaust steam from an engine will inevitably contain a trace of the lubricant employed for the cylinders and valves. It is most important that this oil is *not* allowed to get into the boiler as, if it does, it will cause a great deal of trouble.

To prevent this oil from being carried over into the feed water, the most likely device to be found is a grease separator in the exhaust pipe. This is a globe-shaped casting, flange mounted horizontally in the exhaust-pipe circuit. It has a drain valve at the lowest point. Internally there is a vane of spiral form at the inlet; the outlet pipe is extended somewhat with a baffle across its end, preventing direct escape of the admitted steam.

In operation, the grease separator is automatic in action requiring only periodic emptying of the residue from the drain. The incoming steam has a swirling motion imparted to it by the spiral vane, the effect of which is to throw particles of grease (emulsified oil) and any dirt to the wall of the separator. This now drains to the bottom, while the clean steam gains its exit via the baffled outlet of the separator and then travels to the condenser.

One other condenser worthy of mention is that used on the earliest engines where the exhaust steam entered a chamber and was subjected to a jet of cold water to produce condensation. A vacuum pump was employed as described previously. This simple jet-type condenser may be of use when the original type is no longer serviceable, or is missing, and the restorer wants to provide something which may not need to deal with steam quantities of the order needed when the plant was operating under its original commercial conditions.

Chapter 14
Valves: Types and Service

To get water and liquid fuel to the places where they ought to be, and to keep them out of the places where they ought not to be, various manually-operated valves are needed. (See Fig. 14-1.)

With a few exceptions, these are rotary in operation and maybe a half, or even a quarter, turn from open to shut, or maybe a dozen or more complete turns, depending upon design and function of the particular valve.

Those needing fine adjustment—fuel and possibly water feeds, for example—have many turns to allow for a gradual opening; whereas those that merely need to be open or shut, often have just a half or quarter-turn. These valves include blow-down (See Fig. 14-2.) fuel and water isolating or changeover cocks. (See also Fig. 11-1.)

Quick-action valves are usually of the rotary-barrel type, in which a ported cylinder or barrel rotates in a housing to provide connections with ports in the said housing or, when shut, to give no connections. (See also Fig. 5-2.)

The barrel sides may be parallel or of tapered form. If the parallel type is worn, there is nothing that can be done except either to build up the barrel by electroplating or depositing, or to turn up a new barrel. The taper type can be ground in to renew the seating, using fine grinding paste which must be carefully washed away afterwards. This type of barrel is usually retained in its housing by some form of spring or spring washer with a nut or setscrew in the thinnest end.

After regrinding, check that the end of the taper does not protrude too far out of the housing, thus preventing the spring from keeping it tightly in the valve body. If this does occur, the end of the taper must be

Fig. 14-1. Collection of accessories used in an early steam buggy: (From left to right, top row) steam automatic, throttle valve, mechanical water pump and manual water pump; (second now) changeover valve and oil pump; (third row) hand valve, mechanical pumps operating linkage and manual pump operating linkage, safety valve and fuel automatic; (fifth row) two hand valves, manual fuel pump, fuel filter and mechanical fuel pump; (bottom row) pilot fuel gauge, steam gauges, oil sight feed indicator, stop cock and pressure balance tank.

turned back at least flush with the end of the housing or, preferably, a little further.

Occasionally this type of valve may have a packing gland that needs attention, but this is more likely on the parallel-sided type rather than on the taper valve. Sometimes these cocks have a loose handle that is removed after the valve has been operated.

Multi-Turn Valves

For fine adjustment, a multi-turn valve is usually used which, for high pressure use, normally employs a taper valve in a taper seat of some sort. The simplest form is one of the most common in use and uses the handle spindle as the valve taper, the actual construction being a screwed thread rolled on the shaft where it enters the valve body, mating with a female thread in this body and, at its inner end, being turned to a taper point which seats in a matching taper hole in the base of the valve body. The inlet and outlet passages to the valve are so arranged that one is before and one after this taper seating. Restoration usually consists of either re-cutting the taper on the shaft, if there is sufficient length, or making a new shaft. The taper seat in the body must then be refaced to match, using a taper reamer. Probably the easiest tool to use for the amateur is either a countersink bit, if this is the correct angle, or an ordinary drill ground to suit.

A more complicated taper-seat valve employs a loose valve which is a floating fit in its seating and is screwed down on the seat by a separate threaded rod above it. Often the valve has a male pin which engages in a female hole in the screwed operating rod, in the manner of the ordinary domestic water tap. Often in these cases the valve is made of bronze and usually has a fairly flat profile for the taper and is much larger in diameter than the one-piece rod type. It is seldom necessary to replace the operating rod, but often the valves and seats have to be refaced. If the valves then become too thin, new ones can be turned up from bronze stock on the lathe.

To stop leakage of the fluid past the screw thread, a gland is usually provided. (See Fig. 14-3.) The most common form of this is a packing of P.T.F.E. or graphited string wound around the shaft just outside the

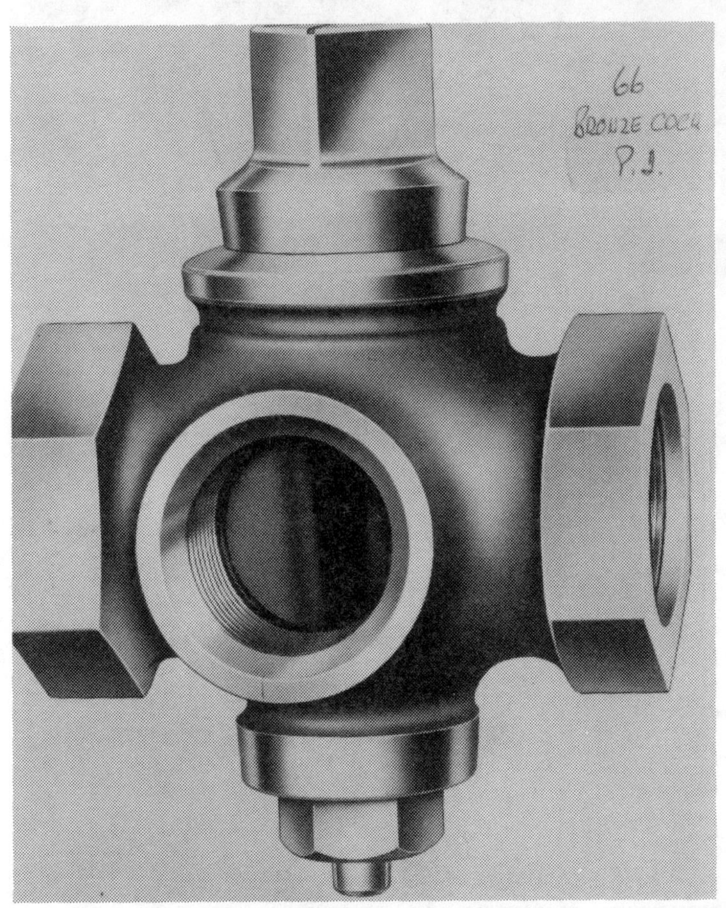

Fig. 14-2. A typical bronze three-way cock was often used as a boiler blow-down valve. Where a lower connection on the boiler feeds the bottom of the water gauge via such a cock, it can be operated to shut off the gauge and open to atmosphere, thus allowing water and sediment to be blown out.

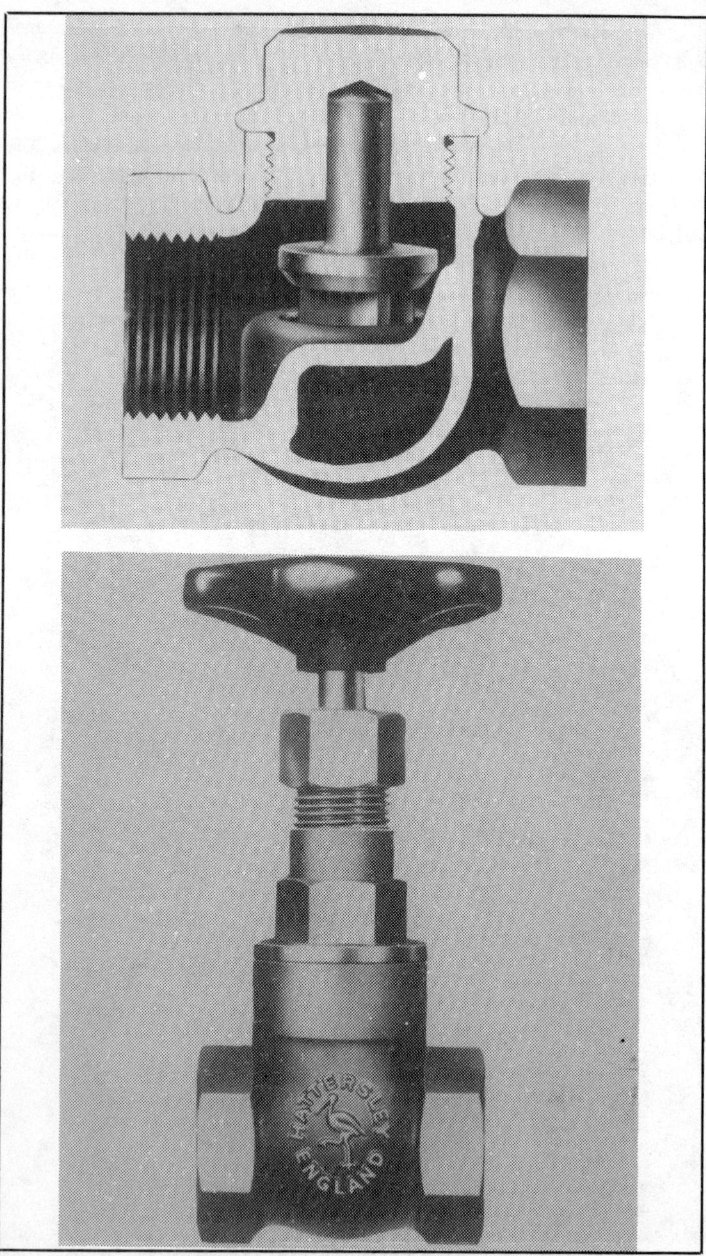

Fig. 14-3. A typical stop cock and a section showing the interior of a similar valve and its loose valve with a taper or mushroom seating—below the hand wheel can be seen the gland nut under which and around the spindle is the packing material to prevent leakage.

thread and retained in position by a cap nut and small brass bush. This nut screws down onto a thread formed on the outside of the valve body.

Repack this gland if the original packing is missing or hard, and then screw down the cap nut just enough to stop leakage. If too tight, you will be unable to turn the valve spindle. If the shaft is badly worn where it rubs on this packing, a new shaft will have to be turned up from stainless steel.

It is better to have the pressure side of the pipework connected to the union under the valve seat. When closed, there is no pressure on the packing gland. In the case of the loose valve type it is essential to connect this way around. If the pressure line is connected to the union above the valve, the pressure will hold the valve on its seat and, thus, shut, even if the operating rod is unscrewed.

Valve Handles

There are a number of ways of fixing the handles to the shafts. An easily removed handle is an advantage for maintenance work on the valve. The easiest type to find, or to make a matching shaft to suit, is the square. It is comparatively simple to file a square on the operating shaft and use a center-tapped hole to retain the handle on the square by means of a setscrew.

A number of different designs of handles have been used through the years by various makers. The so-called "wire-type" consists of many loops of fine wire fitted into a central boss and was used in several early buggy-type steam cars, particularly the Stanley, before they progressed

Fig. 14-4. Controls of an early steam car, the heavy-duty boiler water level gauge and mirror enabling the driver to keep an eye on the water level are shown. Stop cock under the body is a cylinder drain valve. Tall handle is for hand operation of the pumps, pedal is the foot brake and chain is for sprag.

to the cast aluminum type of knob. Other makes used cast iron, either flat or dished. (See Fig. 14-4.)

Gate Valves

Another type of hand valve is the gate type, so-called because a partition, or "gate," drops down to shut off one end from the other. This is a relatively slow-acting valve, usually operated in a similar manner to the taper seat type, but does not give fine adjustment. It is used more as an isolating or back-up valve than for variable supply control. Its merit is that it is strong and does not wear easily; however, when it is worn it is very difficult to renovate, replacement being the recommended procedure. Usually it does not matter which is the inlet or outlet connection.

Some valves with taper seats use a ball bearing to close the outlet rather than a taper rod or valve. Restoration of this type is similar to that described previously except that there is no taper on the spindle, and it is much easier to replace a ball bearing than to re-cut a taper on either a shaft or loose valve. The rod pushes the ball bearing onto its seat, usually having a concave depression in its tip to locate the ball. In this case, for the same reasons as with the loose taper valve, the pressure line must be connected under the valve seat. This type of valve will also work to a limited extent as a non-return valve or check valve, being adjustable for the amount of lift of the ball when used for this purpose.

Check Valves

Although not strictly hand valves as apart from the hybrid type just described, check valves are not usually adjustable by hand. The object of a check valve is to stop fluid from flowing back the wrong way in a circuit, and they are usually used on the water feeds to a boiler and sometimes on the burner fuel feeds.

If you are trying to push water from a water tank at atmospheric pressure into a boiler at, maybe, several hundred pounds pressure, you must have a non-return valve in the system. Otherwise, the pump will refill with the boiler water on its suction stroke, rather than drawing fresh water from the tank.

Many types of pumps have check valves on their inlet and outlet orifices, but you should never rely on the outlet valve of the pump alone. If you do, it prevents you from clearing a blockage in this valve while the boiler is under pressure. Also, if the pipe between pump and boiler, or a union in the line, comes undone, there is nothing to stop the boiler from forcing out scalding water and steam at several hundred pounds pressure.

Therefore, as a safeguard, fit a check valve, or better still a double-check, as close to the water entry point to the boiler as is practical. A double-check is two check valves in series, usually as one unit.

The three types of check valves most commonly found are the *ball bearing type*, the *taper valve*, and the *disc valve*. The ball check valve is

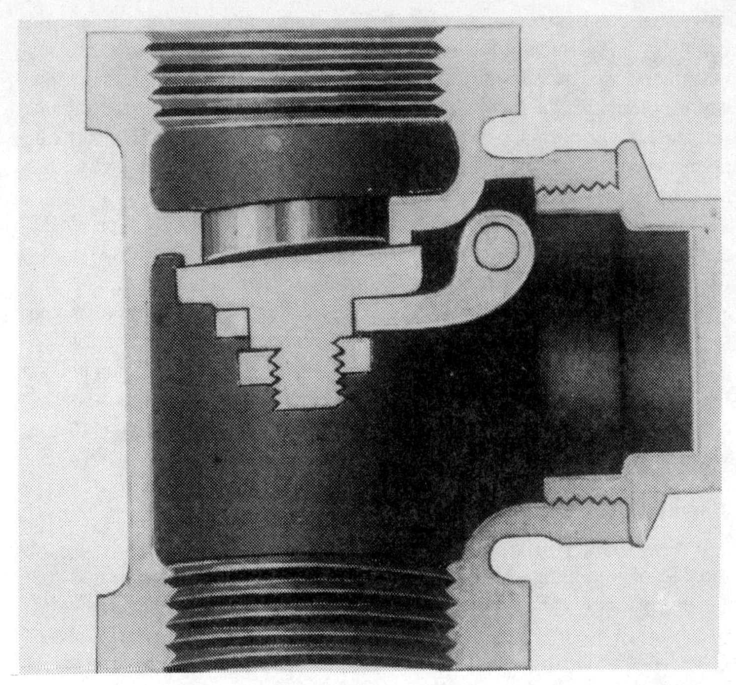

Fig. 14-5. A simple flap-type non-return or check valve, low pressure type through the removable cap, it may be possible to reface the seat if it becomes scored, as well as removing and then refacing the valve.

merely a ball bearing held onto a spherical or conical seat by a light spring or, occasionally, just by gravity and back pressure from the liquid on top. The taper valve can be similiarly located, being a cone-shaped device rather than a ball bearing. The disc is simply that—a flat, round disc that closes the orifice.

The taper and, particularly, the disc valves sometimes are hinged on one side and swing open and shut. They usually close by gravity and reverse fluid pressure without the aid of a spring. These are normally used only for comparatively low pressures.

Renovation consists of refacing taper and disc valve faces, re-cutting the seats in all types (See Fig. 14-5.), and renewing the ball bearings of the ball-type. Replace any springs that appear unsound, either damaged or corroded. Use stainless steel ball bearings and bronze taper or disc valves. Use brass or bronze hinge pins. If you cannot get stainless steel ball bearings, brass or bronze is the next beat.

Occasionally you may come across a multi-purpose valve comprising part check valve and part ordinary open and shut capability of the screw-down type. These are sometimes used for cross feeding between different tanks and their pumps, either for water or fuel, or for by-passing a failed piece of equipment. For example, the author has a car with such a valve that can be used to by-pass the low water automatic, if this should jam shut, and a second for use in switching between fuel tanks. If your machinery does have such valves to isolate safety devices, such as the low water automatic, they should only be used when absolutely essential and returned to the normal position as soon as repair of the defective part can be made. While the particular safety device is thus by-passed, extra care must be taken with operation.

Another odd valve sometimes encountered is the screw-type with a detachable handle. Usually these are used in situations where the ability to adjust the flow of fluid through a particular valve is required, but such adjustment is to be made only occasionally and then not altered for some time thereafter. For example, they are used when setting up for a particular fuel and, thus, probably will not need to be altered again unless the fuel is changed. The most useful type of handle for this purpose is the old-fashioned wash-boiler type—that is, a square hole to mate with the valve shank to which is attached a short, straight handle.

Throttle Valves

A very specialized type of valve is the throttle valve. This has to be capable of operating with infinite variability between full open and full closed, instantly, and to withstand full steam pressure. A slide valve of circular type is frequently used for this purpose. The slide or barrel of the valve has ports in it which mate with the inlet and outlet ports of the valve body, and as the valve moves towards shut, the port area uncovered gradually diminishes. (See Fig. 14-1.)

If the ports are damaged or the barrel badly scored, about the only remedy is to make a new slide. If the damaged ports are in the valve body, the body will have to be reamed out to clear the damage and an oversized new barrel made. The same applies if either body or barrel is scored. Fortunately, however, such damage or scoring is rare and, with luck, the only work needed will be the repacking of the gland.

The gland nut must not be tightened down too much or the valve will jam and a sticking throttle valve is one problem you can well do without. Do not forget that, in most cases, you cannot de-clutch a steam engine from the driven machinery. Therefore, if you cannot shut the throttle, you are quite likely to end up with bits of machinery in all directions.

If the original valve or its handle mechanism is missing, it will have to be acquired or made. Try to find another example to copy. Failing this, make up a suitable throttle valve. Then devise the linkage to the control lever, be it hand or foot, so that there is a reasonable movement between open and shut. Ease of operation should be the main criterion.

A previous remark is hereby repeated, namely that it is wise to have a screw-type back-up valve between boiler and throttle valve that can be shut should the throttle fail for any reason. It can also be used to safeguard unauthorized use of the equipment, if it is left with steam up at any time, unattended.

Chapter 15
Pipework
Fittings and Threads

Pipes are bits of tube that convey liquids or gases to and from places where you want them to be and, hopefully, keep them away from places where they are not required.

Plastic pipe has no place in steam appliances, thus the materials most likely to be used are steel and copper. The authors recommend for all steam plant pipework, no matter if it is in fact carrying steam or simply water and fuel, that you use high-pressure, solid-drawn steel or copper tube.

All steel pipe should be threaded and joints caulked. Use liquid or paste jointing and, if necessary, string or tow. A suitable jointing is Boss White, or a similar substance.

Copper pipe should be threaded and the unions soft-soldered on the threads, or if non-threaded unions are used, these should be silver-soldered to the pipes, *not* soft-soldered. Except for the low-pressure side of the water pumps, that is from tank to pumps, never ever use domestic-type compression fittings. Such joints, if properly made, can withstand considerable pressure, but it is difficult to tell if the compression ring has crimped down into the pipe really tightly, thus the recommendation that the amateur should not use this type of fitting, except for the low-pressure work previously mentioned. If you do have to use such a fitting in any other location, get a qualified plumber to either make the joint or check your work.

Unions and fittings for steel pipe should be of the forged-type and for copper pipe, solid brass. (See Fig. 15-1.) Watch for any sign of these splitting when tightening up as this sometimes happens, especially when tightened onto a taper thread.

Pipe Sizes

The sizes of pipe depend upon the size of your steam plant and the various applications. For most restoration projects that the amateur is likely to attempt, pipes in the range between 1/8-inch and 1-inch bore are all that will normally be used. As a general rule, but not invariably, the larger sizes will be in steel and will be used for steam, whereas the smaller pipes will be copper to carry liquids, water and fuel, and gas fuel.

Here is an example of the average small steam buggy, or about a 10 hp stationary plant: The steam mains would probably be 5/8-inch to 1/2-inch bore steel pipe, with maybe a form of swivel or flexible coupling to the actual engine, which will be dealt with in more detail later. The blow-down or drain-valve pipes will probably also be the same size and material.

Water feed pipes from tank to pumps will suffice at 1/2-inch either steel or copper; from pumps to boiler, 3/8-inch copper or steel, copper being possibly better. Fuel lines are likely to be, and it is recommended that if they have to be renewed they become, copper, maybe 5/16-inch or 1/4-inch bore for the main jet and pump feed circuits and down to 1/8-inch for the pilot feed. It may also be advantageous to reduce the diameter of the main jet pipes near the jet, especially after the vaporizer. The smaller the pipe size, the faster the fuel-speed for a given quantity per minute, thus, the less time to cool off again after being vaporized. This must be balanced against the fact that a larger copper area provides more heat retention.

When working out the connections between the various pieces of equipment, it should be borne in mind that you may wish to remove any one piece for replacement or repair, and the ability to do this without disturbing anything else is most useful. The connections should be made in such a way that there are suitable disconnecting unions or points for each pipe leading to or from the various items. Sometimes there will be a nipple and union locking nut securing the pipes, which is fine; but, in other cases, the pipe may screw directly into the equipment, making it necessary to incorporate a line-type disconnecting union in the pipe. Where there is a pressure line feeding any particular item which it is likely you may need to remove, it is a good idea to have a stop cock on that pressure line so that the pressure can be retained when the faulty piece of equipment is removed.

Pipe Fittings

The types of fittings most useful are: elbows, tees, bushes, locknuts, disconnecting unions, nipples, gradual bends, crosses, double-bends, plugs, flanges, and sockets. In forged-steel fittings, there are a variety of types for most of the items. (See Fig. 15-2.)

Elbows come in male/male, male/female, and female/female sorts, with angles ranging from 45 to 180 degrees. Tees, again, come in a selection of male and female types with variations on their angles to give "Y," or 45 degree, branches, as well as the more normal right angle. Nipples are all male/male, but the two ends may be the same size or of reducing type—that is, one side larger than the other. Plugs come in solid or hollow versions, with raised or countersunk heads. Bushes can be parallel or taper reducing male/female types. Bends come in all angles and varieties of male/female ends. Sockets can be parallel or taper thread, or reducing, and female/female or male/female. Disconnecting unions can have flat, spherical or conical seats. The conical seat is advised except where the two pipes cannot be exactly lined up, when a spherical seat is useful. The flat-seat type is not recommended except for low-pressure work.

Many of these fittings are also available in brass or bronze, but the range is more limited in these materials.

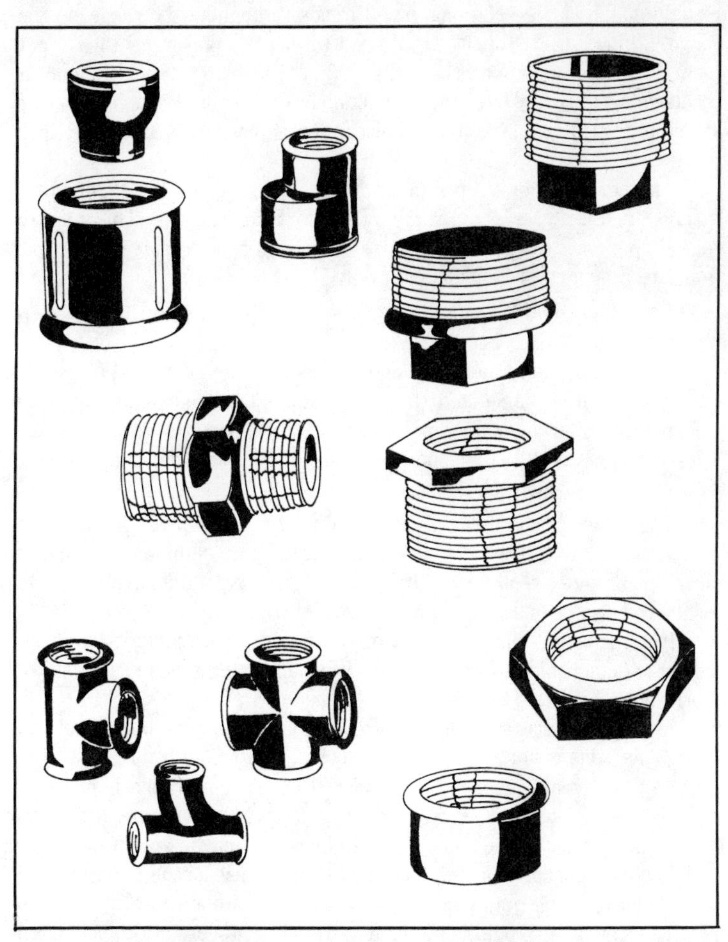

Fig. 15-1. A selection of pipe fittings showing from top left: sockets and a nipple; two types of plug; a bush, lock nut and cap nut; two types of tee and a cross. These fittings usually have the Standard Pipe threads and are made in diameters upwards from ⅜-inch.

Pipe Threads

One of the main problems you are likely to encounter is the multitude of threads that have been used down through the years. It does not follow that because the plant has been manufactured in a particular country that it necessarily uses entirely the standard threads of that country.

Some threads are so nearly the same that in most cases they are interchangeable. For instance: U.S.A. National Coarse, British Standard Whitworth and Unified Coarse are all very similar, the differences that do exist being mainly those of thread form or shape, rather than pitch.

British Standard Pipe threads, both parallel and taper forms, do not exactly match U.S.A. National pipe threads throughout the range. In the range 1/8-inch to 3/8-inch, the British thread has one more thread per inch than the U.S. counterpart. For the 1/2-inch to 7/8-inch diameter inclusive sector, they have the same number of threads per inch, and from 1-inch to 2-inch inclusive, the American thread has a half thread per inch more.

Another commonly encountered thread, especially on small brass fittings, is the "gas," or British Standard Brass thread. This has 26 threads per inch throughout the size range and approximates to British Standard Fine for 1/4-inch diameter only.

Then there is the Unified Fine series, and a whole host of different metric types.

You will certainly need a range of taps and dies to thread pipework and fittings. It is suggested that, as a start, taps and dies should be obtained for pipe diameters in the range 1/8-inch to 3/4-inch inclusive. Preferably, get both British and American pipe thread types and 1/8-inch, 1/4-inch and 1/2-inch "Gas" sizes.

These "gas"-type dies will be useful for threading pipe to mate with fittings having this thread, which may well be quite common, depending on the plant under restoration. If available, the type of die having a guide for the rod or pipe is useful to enable a concentric thread to be cut.

In some cases, there may not be room to use such a guide, so make sure it is detachable before purchase. When cutting threads use plenty of lubricant as this helps to prevent a thread from stripping.

In most cases a taper thread is required on the pipes, both steel and copper, as this enables them to be screwed into the fitting until tight enough to prevent leakage. A reminder: Use paste or liquid jointing on steel pipes, and soft-solder the joint after screwing in with brass fittings and copper pipes. In the case of steel pipework, maybe an elbow has to end up pointing a particular way. You will have to use a running or parallel thread, with a lock nut, wrapping string or tow around the thread with liquid jointing before pulling up the nut. This is a process not to be used if it can be avoided.

Another *do not,* if you can avoid it, is the joining of steel to copper, as electrolytic action between two dissimilar metals causes corrosion. In certain cases, such as water feed to the boiler, it may be impossible to do anything else.

Connections

When connecting up, especially in confined spaces such as usually exist in the steam buggy, it is best to first install the steel piping, as this is not so easy to bend. In fact, without a proper tube bender, such as a Hilmor, it is almost impossible so to do neatly. In most cases, bends and elbows are all that is needed. If you must bend a length of pipe, make a template from thick wire, take the pipe and template to your plumber and ask him to do it.

A trial run, assembling all joints dry, and not absolutely dead tight, is recommended. You will often find fittings screw up far more on taper threads than you imagine, so that you can then check that the lengths will be correct. You can also get an indication if any fitting will not turn far enough on any particular pipe to end up pointing in the right direction. It may take you several trial attempts before you get everything neat and pointing the correct way.

With one exception, which will be mentioned later, one does not normally want any lengths of pipe to form a liquid trap—that is, in effect, the bottom of a "U." In the case of water, any such trap prevents complete draining and could freeze and burst in very cold weather.

The fuel lines, and may be water feed pipes, are often carried out in copper, as mentioned previously. This should be annealed—that is, heated to a cherry red and quickly quenched in cold water, making it soft and easy to bend nearly to follow the required route. These pipes can usually be quite easily threaded around any previously installed steel piping or equipment. Remember that if holes are needed through any partitions or

Fig. 15-2. Selection of fittings by G. Slingsby of Hull, showing bends, tees, couplings and bushes, an expansion coil and a pressure gauge. A good supply of such fittings will be needed for any major restoration work on a steam plant.

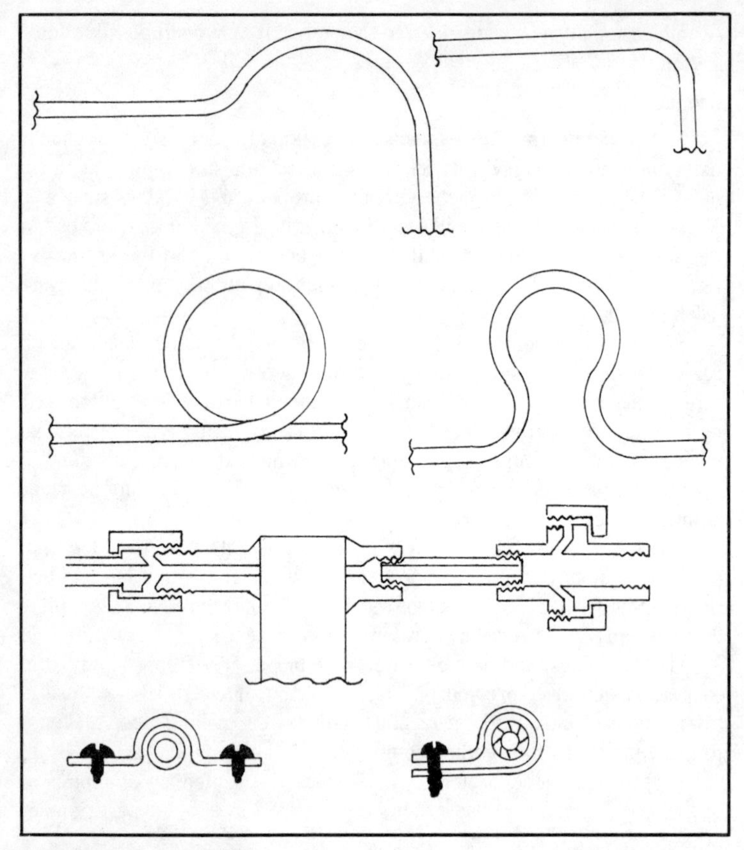

Fig. 15-3. Pipework details.

parts of the bodywork, in the case of a road vehicle, these holes should be big enough to accommodate any union nuts that may be on the pipes.

Try to make all bends neat and not too sharp. Sudden elbows not only look bad, but they also restrict flow. If a quick bend is needed, use a custom-made elbow. It both looks better and, due to internal radiusing, works better. (See Fig. 15-3.)

Loops and Flexible Joints

Where any copper pipes are subject to expansion due to heat or vibration, incorporate a 360 degree loop in the line to absorb same. (See Fig. 15-3.) Steel pipes generally do not expand enough to cause any problems and are much more rigid to withstand vibration, thus, expansion or vibration loops are rare. With all pipework, and in particular with copper, clip neatly to the frame or bodywork at frequent intervals. This not only looks better but also prevents sagging, chafing and, to an extent, vibration. Use steel clips for steel pipes and copper clips for copper pipes, because of the electrolytic action mentioned previously.

The loops should be above the pipe for drainage purposes, except in the case of the line carrying steam to the steam gauge. This is the one aforementioned exception where the loops is below the line. The reason for this is that by having a loop below the line, steam will tend to condense out to water therein and be trapped in the bottom of the loop, thus forming a barrier between the steam and the gauge, applying pressure by a liquid, rather than a vapor.

When incorporating taps with fixed single-arm handles, try to arrange them so that if the handle turns due to vibration, it will drop down to the normal operating position. For example, on an oil feed tank, if the tap should move during use, you want to ensure that it moves to the "on" position, not to the "off."

To transmit the steam from the throttle valve to the engine, it is sometimes necessary to incorporate a flexible joint. This particularly applies in any application where the engine is free to move relative to the supply pipe. For example, some steam buggies have the engine direct-coupled to the rear axle and, thus, move up and down to a certain extent with the springs.

The three most common devices used for this purpose are steam tight, swivel joint, a large gentle curve "C" pipe or a high pressure flexible hose. Flexible pipes, in time, break;thus, it is prudent to have a spare readily available.

Lag all steam pipes where they are outside the boiler and the water feed pipe, if any of the section after the feed water heater is outside the boiler. Asbestos tape is the most convenient material for this purpose. Wrap around and around the pipe, finally tying-off with copper wire.

Chapter 16
Care and Repair of Gauges

The pressure gauge probably springs to mind as being a typical gauge in that it has a round dial with figures and a pointer or needle pivoted in the center, the whole device being enclosed in a case with a glass front. This case was usually made of brass in steam plants and was always kept highly polished or in some cases—particularly steam buggies—nickel plated. The gauge could be either mounted on a flat surface or recessed into a panel, as in a car dashboard. (See Fig. 16-1.) In either situation, the casing of the gauge usually had a flange incorporated to enable it to be screwed or bolted in the desired location. (See also Fig. 14-4.)

Bourdon Tube

The type of gauge used for registering boiler pressure or steam pressure in a pipe or engine is the Bourdon-tube type. (See Fig. 16-2.) The essential part of the mechanism is this Bourdon tube, which is a very thin-walled tube usually of oval cross section and made of phosphor-bronze. This is curved to form three-quarters of a circle, one end being fixed into the brass block forming the inlet connection. The other end of the tube is plugged and provided with a connecting pin projecting from it which engages in a light linkage pivoted in such a way that a rack sector on the other end of this link engages with a pinion fixed to a thin shaft which is able to rotate in brass trunions.

The pointer or needle is mounted on one end of this spindle, under which is the dial face. In operation, pressure in the tube tends to straighten the oval tube, and as one end is fixed, the other moves the linkage and rotates the pointer. Fortunately for the simplicity of the gauge, the amount of movement of the oval tube is directly proportional to

Fig. 16-1. Dashboard of Twenties period Stanley steam car—note the rod standing vertically in front of the seat is not the hand-brake, but is for pumping fuel and water by hand. Left-hand pedal is reverse; right-hand is brake. Throttle is the lever under the steering wheel.

the pressure applied, thus the pointer will move in constant relation to the pressure and a linear dial marking can be used. Around the spindle is a light hair spring which takes up slack in the linkage and helps to return the mechanism to a zero reading when the pressure is removed.

There is not really much that can be done in the way of repair or service to this type of gauge, as the calibration has to be expertly carried out to ensure accurate readings. Fortunately, the repair of such a gauge by a specialist firm is not likely to be too expensive and should not be too difficult to locate, as most restorers of gasoline car gauges could tackle such work.

Two jobs can be done by the amateur if he likes and feels he has the ability. First, the easy one is to renew the glass. Second, and that which requires good eyesight and a steady hand, is re-lettering and numbering of the dial face. If the face is made of paper or thin card, this writing must be done in Indian ink; if of metal, then oil bound paint should be used.

Gauges of this type are also suitable for oil, water or fuel pressure pipes or tanks.

When required to indicate steam pressure, the gauge must never be mounted directly in communication with the steam, but on a siphon. This is a U-shaped or single coil of tube between the gauge and the boiler or other steam source. A certain amount of steam condenses in the "U" or lower part of the coil. This has the effect of insulating the delicate Bourdon tube from the heat.

A gauge required to indicate vacuum or negative pressure, for chimney "pull" for example, is exactly the same in principle; but, in this case, the arc of the Bourdon tube tends to reduce in diameter rather than straighten out, and the sector rack and pinion mechanism that turns the needle is arranged to give the appropriate reading on the dial.

If a replacement gauge has to be fitted, choose one with a range such that the normal pressure is indicated about halfway. For example, a

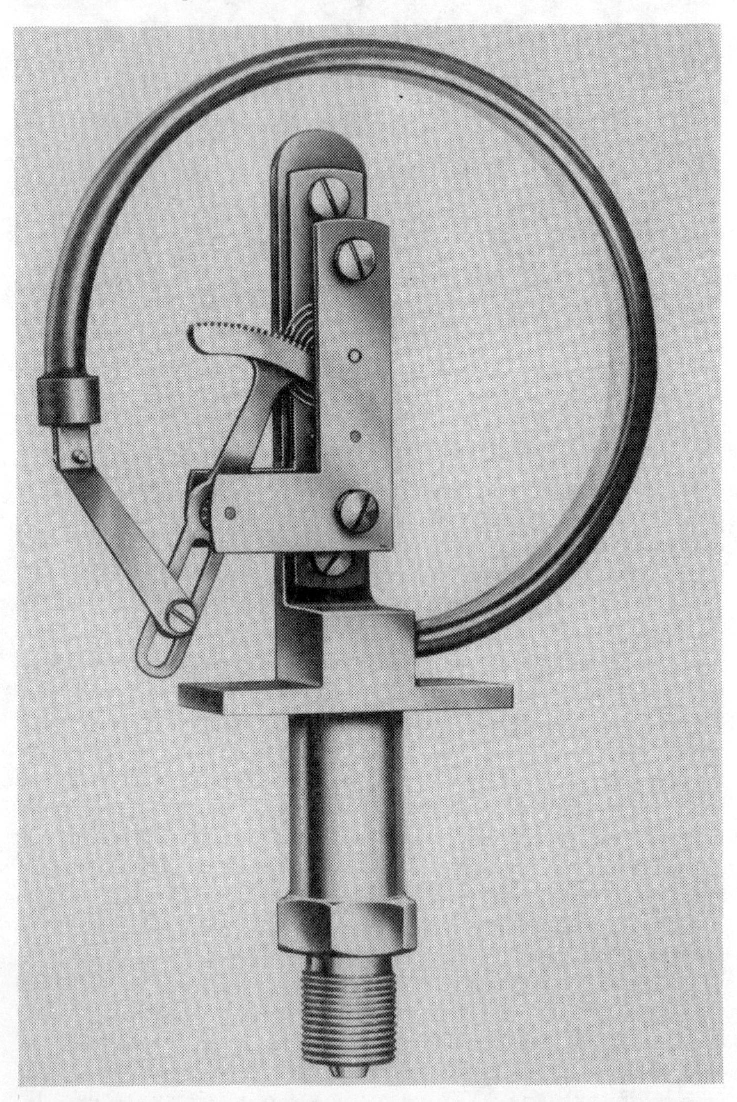

Fig. 16-2. The interior of a Bourden-type pressure gauge—the hollow oval, or almost flat, section tube tries to unwind as pressure is applied to it, causing the sector lever to rotate on its center pivot, in the process turning the center spindle to which the gauge pointer is fitted.

working pressure of 160 p.s.i. is best indicated on a gauge with a range 0 to 300 p.s.i. This will avoid the possibility of straining the mechanism.

You may encounter a pressure gauge which works by the deflection of a diaphragm. The characteristic that indicates this is the horizontal circular housing under the body. The same comments as previously discussed apply to the servicing of this type; they are more often employed for the measurement of pressure in liquids.

Contents Gauge

Another application of a pressure gauge is that of indicating the contents of a tank. The pressure range is very small and the gauge is very sensitive, as the head of liquid available is not usually more than about four feet. The dial is marked in units of quantity (i.e., gallons or liters).

The more usual tank contents gauge is a transparent tube of some sort, mounted vertically adjacent to the tank, enabling a direct reading of the level to be obtained. This is a rather more simple design than the similar type used as boiler water level gauge, as the pressures involved are very low.

The top and bottom fittings usually incorporate screw-down or gate valves, to enable a broken tube to be isolated and replaced. The glass-tube type can incorporate gland-type seals on the tube (See Fig. 16-3.) or, in the case of short glasses, the tube may simply be bedded between sealing washers at each end. Replacement washers should be of a material suitable for the liquid being measured. Rubber is suitable for a water tank but not for any mineral-based liquid, for which purpose Neoprene or an oil-resistant material should be used.

Tank contents gauges should have some form of protection against accidental breakage of the tube, particularly the glass type. These may be quite simple, such as two or three vertical rods between top and bottom flanges, or a brass tube with a vertical slot, enclosing the transparent tube. The heavy-duty protectors required on boiler water gauges are not needed.

Advantage may be taken of modern materials in this application. For example, use may be made of plastic tubing, such as silicone, which is resistant to the attack of the sort of liquids likely to be encountered with steam plants. This tube is sufficiently transparent for this purpose, is virtually unbreakable and is flexible. Flexibility is a great bonus as it means that critical alignment of the top and bottom fittings is not essential, as it is with glass (as the authors know to their cost). In fact, top and bottom fittings do not have to be even in line and, in extreme cases, can be in a different plane.

This application is the only one where we would suggest that you might consider using a plastic pipe or tube on a steam plant. Despite the ease of routing and fixing, *do not be tempted to use plastic pipes or tubes on steam work, or for any other part of the system.* Even when used for these particular gauge glasses, keep well away from any heat.

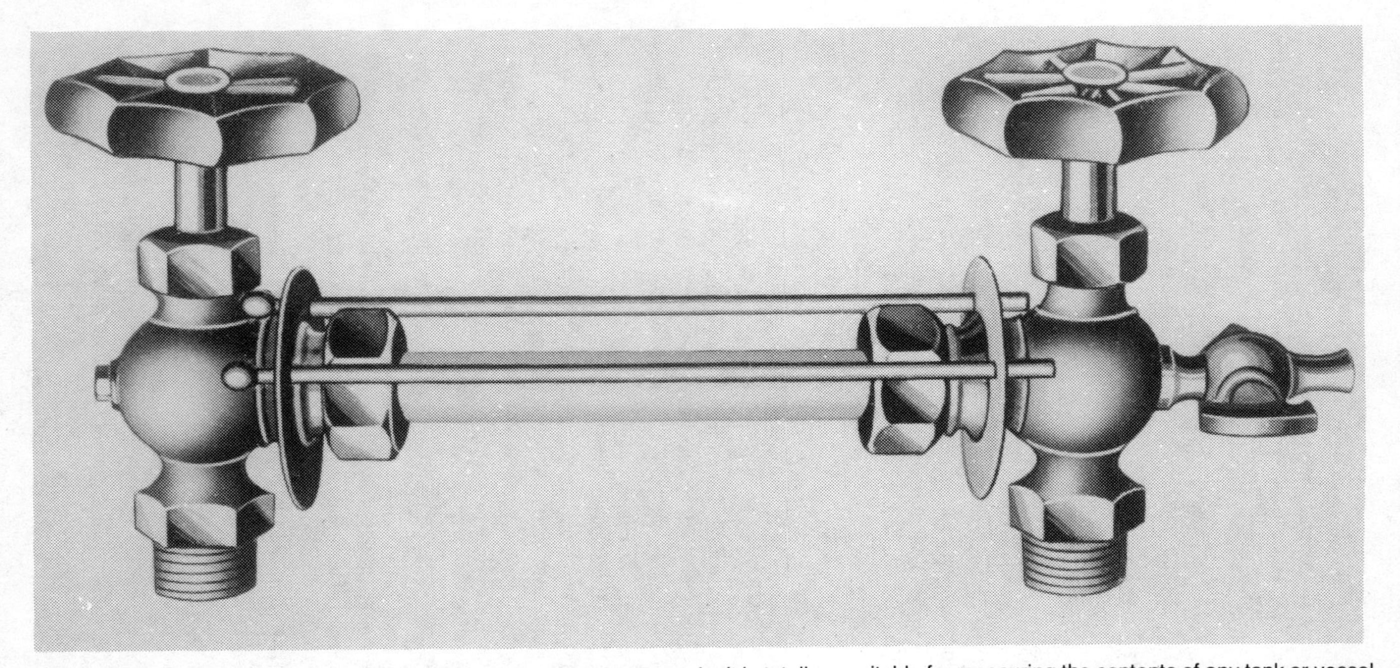

Fig. 16-3. A fluid level gauge is most definitely a low pressure gauge only. It is totally unsuitable for measuring the contents of any tank or vessel to which pressure other than atmopshere or gravity is applied. There are glands at the top and bottom of the glass tube which contain packing material. If the union nuts are overtightened, the glass will break.

All gauges of any type should be mounted in a position as free from vibration as possible. This will help to preserve the accuracy of the instrument for as long a period as possible. In any event, pressure gauges must be checked regularly against a test gauge of known accuracy.

Oil-Winkler

Another unit which might be termed a gauge, but which is a bit odd, is the oil sight flow indicator, sometimes called an "oil-winkler." The object of this is to provide a visual check that the cylinders are receiving a supply of oil; thus, the gauges are situated in the oil line between pump and engine.

There are various forms that these may take. Most are quite small, only a few inches in overall length, and may or may not have moving parts. Those with moving parts usually have a vane or some form of rotor which is supposed to spin as the oil passes through. The type without any moving parts relies on the passage of the oil being actually observed as it wends its way along a winding track or pops up, or "winkles," out of a small central orifice into the surrounding case.

All types have at least one glass side which must be made of thick glass to withstand the considerable pressure. If damaged or scratched, or in any way not clear, this glass must be replaced. Explain to your glass man exactly for what purpose you want the glass, and make sure he realizes the pressure it must withstand.

For oil to enter the engine cylinders it must be above steam working pressure, otherwise it obviously will not go in! Usually, the gauge glass is flat and, thus, no problem to replace. If it is of the domed type, however, about your only possible source of supply is the original manufacturer. Failing this, the unit will have to be renewed *in toto.*

With heavy, dark, steam-cylinder oil, often all you can see in these gauges is oil or no oil, rather than any actual flow. It is, therefore, helpful to keep one eye on the actual pump to see that it is working. This is described in the following chapter.

Chapter 17 Lubricators and Pumps

The supply of the right grade of lubricant in the right place, at the right time, and in the right quantity, is what lubrication is all about and is vital to the trouble-free running of any machine.

The movement or lack of movement of the part to be lubricated affects, to a certain degree, the method of lubrication, as do the temperature and pressure at which the part normally operates.

For the purposes of this section, we can conveniently deal with lubrication methods in two parts—first with cylinders and valves and then with the rest of the mechanical parts.

The first part, the lubrication of cylinders and valves, concerns parts which are subject to reciprocating movement of a sliding nature, high temperatures and steam pressure. All this demands the use of a thick oil and an appropriate means of maintaining a reliable supply.

Forced Feed Pumps

Probably the most reliable and widely used type is the mechanically-driven, force-pump lubricator which has, in fact, come to be referred to as simply a cylinder lubricator. Two very well-known patterns are the "Wakefield" and the "Silvertown" types, the operations of which are similar. The body consists of a cast iron box with a hinged lid and lugs or feet by which it may be affixed to or near the engine.

Running through the body near the top is a shaft on which are fixed a number of cams. Below each cam is a plunger pump mounted vertically so that the plunger is directly driven by the corresponding cam enclosed within a frame on top of each plunger. Every plunger operates in its own cylinder which has suction and delivery valves incorporated in the base.

These valves are of the ball-type with springs to assist their seating while working in a viscous oil.

The body of the lubricator acts as the oil reservoir so that the inlet valve is constantly fully immersed and, thus, free from air locks. The delivery is by means of an external connection.

When the drive shaft is rotated and a plunger is moving up in its cylinder, oil is drawn in past the inlet valve and fills the space below the plunger. Further rotation of the shaft causes the plunger to descend, forcing the oil past the delivery valve and through suitable pipework to the engine cylinder.

This type of lubricator seldom gives trouble, as all of its working parts are continuously bathed in fresh oil in an enclosed housing. There are cases of examples working quite well for 50 or more years without any trouble.

An adjusting screw is provided to regulate the quantity of oil delivered. This screw usually limits the amount the plunger can rise up the cylinder, thereby altering the effective length of the stroke. As a rough guide on a vehicle, the lubricator should be adjusted to deliver from one to two ounces of oil per 100 miles, depending upon the size of the engine.

The drive shaft is rotated by linkage taken from some convenient part of the engine and, since this usually means a reciprocating motion, this is transferred to a rotary action operate a pawl which rotates a ratchet wheel, either directly on or connected by gears to the lubricator shaft.

The most likely fault with a lubricator of this type fitted to an engine that is being restored is that in all probability it has been lying empty for some time and the contents may be found to consist of rusty water and a greasy sort of mess. Do not despair, as this usually looks worse than it is. After cleaning out the "goo," the innards will, more than likely, be found to be in good condition. Dismantle the pumps and drive mechanism, taking care to keep each plunger matched to its own cylinder. The parts may now be thoroughly cleaned and any sealing washers replaced; these are often leather and will harden without the oil.

The ball valves should be checked for seating and new balls of bronze or stainless steel used if necessary. The ratchet drive may have suffered more, as being outside the body it is not automatically lubricated; wear may be apparent on the toothed wheel or any gears involved, as well as the pawl and its linkage. The gear wheels are unlikely to be so badly worn that they cannot be cleaned up for use, but you may not be so lucky with the ratchet wheel or pawl. In this case, repair or make a new one as described later in this chapter for the plunger-type pump.

The oil delivery from the lubricator should be conveyed in copper pipes to the cylinders. Several outlets may be involved feeding, apart from the cylinder itself, the valve chest, and the piston rod gland. For a two cylinder engine, the lubricator could have at least six plungers. It is essential that a non-return ball valve is fitted at the point where the oil enters the steam space in order to prevent steam and condensation under pressure from backtracking to the lubricator. If this happens, it is possible that steam or water will get into the body past the output valve. The effect of this is to displace the oil therein, thus cutting off the supply.

Hydrostatic Lubricator

This effect is, in fact, made use of in another type of cylinder lubricator which consists of a body having a lid, an oil connection near the top, and a drain cock or plug at the bottom. The oil outlet features a fine screw regulating valve. The operation of this lubricator type, termed "hydrostatic displacement," is very simple. Condensed steam enters the body causing the oil to float on the surface and overflow through the regulating valve and then into the delivery pipe. (See Fig. 17-1.)

Before all the oil is used, the condensate must be drained off and the lubricator refilled with fresh oil. Some types have the oil delivery passing through a sight-glass filled with water or brine, so that the operator may see that the oil is actually being delivered and, hopefully, at what rate. (See Chapter 16 on gauges.) Two or three drops of oil a minute ascending the sight-glass should be sufficient.

As there are no moving parts, there is virtually nothing to go wrong with this type; but in operation it can have a great disadvantage in that the feed can be intermittent, depending upon whether steam is on or off at the engine. Obviously, the engine can coast without steam being applied and this can mean either no oil feed or a huge gulp of oil due to the pumping or sucking action of the engine under these conditions. This effect is usually more marked with a road vehicle, as it can coast down hills for long periods; whereas when steam is shut off on a stationary plant, usually only the inertia of the engine keeps it running and this only for a comparatively short while.

Plunger Pumps

Another type of cylinder lubricator found on some steam cars is the plunger-type, similar to that used for pumping water or fuel in the system.

There are two types of these but both use a similar pump, basically the same as the above fuel pumps but in miniature size. The high pressure needed to overcome steam pressure is usually obtained by making the pump plungers of very small diameter, while the stroke is made about the same as the other pumps in the system. This enables one drive linkage to operate all pumps quite easily. The Stanley and many other makes used the crosshead to drive these pumps, with various forms of linkage or rocking shafts. (See also Fig. 11-1.)

The variation on this was, in fact, in the drive to the pump. Particularly for the smaller sizes of engines, it was difficult to construct a pump sufficiently robust, yet with a small enough plunger, that a delivery of oil for every movement of the cross shaft did not result in excess oil

Fig. 17-1. A Worthington duplex steam pump, a single-cylinder device, is an ideal restoration project for a small space. This is an example of the displacement-type lubricator on top of the valve chest, together with screwed gland nut type of stuffing boxes.

being delivered. However, the fuel and water pumps did need this frequency of operation.

To overcome this, a ratchet drive was introduced in which a pawl on the driving linkage slowly rotated a ratchet wheel which had, built integral with it, a spiral ramp with a sharp apex. As the ratchet wheel rotated, a small pin attached at right angles to the pump plunger was forced up the spiral until, at its apex, a spring on the plunger shaft slammed it smartly into the pump body, thus "injecting" a measure of oil into the steam pipe near the cylinders and via a non-return valve. (This valve should be located immediately adjacent to the steam pipe.) (See Fig. 17-2.)

Adjustment of the amount of oil delivered was reasonably easy to alter as it merely meant altering the linkage so that the ratchet wheel was rotated more or less at each "bite." There was also usually a means of altering the spring tension on the plunger, if it was required to increase or decrease the pressure of the feed.

These plunger pumps are not likely to give much trouble. Overhaul is virtually identical to that described for the fuel pumps; but it should be emphasized that usually stainless steel is a better material for the plungers than phosphor-bronze, and stainless steel ball bearings are better for the check valves than bronze.

The part most likely to give trouble is the ratchet drive, if used. Often it will be found that all or some of the teeth on the ratchet wheel are badly worn. Since these are normally hardened, they must first be let-down before a file can be used to reshape to somewhere near the original profile. Hopefully, at least some of the original teeth will be in good enough shape to show this. After filing, the teeth must be case hardened by heating to a cherry red and then dipping in "case-it" or similar material. If such is not available, quench in ordinary engine oil.

If the original ratchet wheel is too badly worn to reclaim, or missing, Klaxon horns have an internal part that might well suffice, either obtained

Fig. 17-2. Ratchet-type cylinder oil pump—the peg on the shaft in the center of the picture rides up the ramp as the ratchet wheel is turned until it reaches the ramp crest, at which point the spring on the right shoots the plunger into the oil pump shown at the left-hand side.

new from them or from an old car antique horn. Small boats still use this type of horn, so it is worth trying your local ship chandlers.

The other part of this linkage prone to wear is the pawl. Here you will, in most cases, have to make a new part. This, again, requires hardening; but it should be an easy process as they are not very complicated. The important thing is to make sure you get the tooth of the pawl matching the teeth on the ratchet wheel.

For road vehicles with ratchet-type pumps that are capable of any reasonable turn of speed, one problem that may arise is that gravity return of the pawl onto the ratchet wheel is not good enough at the higher speeds. It may well miss some of the time, rather than taking another bite. A light spring should be added to hold the pawl down onto the ratchet wheel. This, in turn, sometimes produces another problem; now, instead of riding back up the ratchet teeth for a fresh bite, the pawl pushes the ratchet wheel backwards. To counteract this, make some form of spring-loaded, non-return pawl and fit to the other side of the ratchet. Occasionally this can be incorporated into the operating pawl.

Keep the moving parts, including the ratchet wheel and pawls, well oiled. This is usually a matter of going around with an oil can every so often. Also, oil the springs as this will prevent them from going rusty. When rebuilding the pump, make sure the spring on the plunger, if any, is in first-class shape. A poor spring here will prevent the pump from working, no matter how good every thing else is.

General Lubrication

Now the second area of lubrication, those parts external to the cylinders, will be considered. As mentioned earlier, the movement of parts can influence the method of lubrication. A simple pin in a bushed hole forming part of the valve gear may only be subject to intermittent partial rotation. In this case, a hole leading to this pin, through which oil can be fed with an oil can now and then, will suffice.

For a connecting rod, big-end bearing advantage can be taken of its circular motion to provide an automatic feed so long as it is turning. An enclosed box with filling plug is formed on top of the bearing and an internal pipe projecting some way up into the box communicates with the actual bearing surface. As the bearing follows its orbit around the crankshaft, oil is thrown up and enters the pipe on each revolution. To ensure an even feed, the pipe has some sort of restrictor in it. This can be in the form of a grooved pin or a plug of worsted (wool) strands held together with a twisted strand of copper wire. The amount of the feed can be regulated by the size of the pin or the number of strands of worsted.

Another use for worsted-type oil feed is in the siphon lubricator where no movement is available to assist the process, such as main bearings, glands, slidebars and valve gear parts. The construction is similar to that described above, but the worsted plug is must longer and the strands project from the top of the tube and lie in the oil box as "tails."

The oil is then soaked up by the wool and delivered down the tube. This action, once begun, continues until the oil is used up, no matter whether or not the engine is in motion. If you wish to avoid an even more "gooey" mess than is usually found in a steam plant, lift the trimmings out of the siphon tubes at the end of use.

To make a new worsted "plug" trimming for the big-end bearing, or similar application, proceed as follows: First, measure the distance from the top of the siphon pipe to the bearing, using a piece of wire. If, for example, the distance is four inches, the copper wire should be twisted for about two and one-half inches in the middle part, the doubled ends being at the bottom. Wrap the worsted strands lengthwise over the twisted part of the wire, noting how many strands are used in case of the need for future alteration. After wrapping the worsted, twist the ends of the wire at the top to form a circle passing the ends of the trimming through this. The circle of wire should be such that when the plug is inserted in the feed tube, the top will be just under the cover. This serves two purposes: it prevents the wire at the bottom of the tube from actually coming into contact with the bearing, and by touching the underside of the lid, the whole trimming is prevented from being thrown out when the crankpin is revolving.

There should be a small vent hole in the cover in order to prevent a vacuum from forming as the oil is used up, and this hole should be plugged with a cork or piece of cane. Both these materials are slightly porous and allow air to enter but prevent the oil from being thrown out.

On some stationary engines, the lubricator body may be of glass, and some may have a little valve operated by an over-center type lever so that the oil flow may be easily and conveniently turned on or off as required.

In an open-type engine, these lubricators are generally fitted directly onto their own particular bearing but, in the event of these locations not being readily accessible, a larger lubricator with several siphon or feed tubes may be used. This is positioned for ease of filling by the operator and the feeds led to the desired points via copper tubing. Large and powerful engines employ a mechanical lubricator of the type used for cylinders to force feed the oil into the vital areas.

A more simple type of mechanical feed pump is sometimes used on steam buggies very much as used on gasoline cars of the era. In this the driver depresses a plunger in a cylinder and a spring gradually returns this plunger, thus giving a continuous controlled feed. The exact output is adjusted by a tapered screw in an orifice. There are no inlet or outlet valves, the plunger seal performing this task by being cup-shaped and made of a pliable material such as leather. This cup is mounted with the concave side uppermost so that, as it is depressed, oil is able to deflect the lips inward slightly and transfer itself to the upperside of the seal. As the spring returns the plunger, the oil above forces the seal lips into contact with the cylinder walls, forming a tight seal and leaving the delivery pipe as the only way out for the oil. (See Fig. 17-3.)

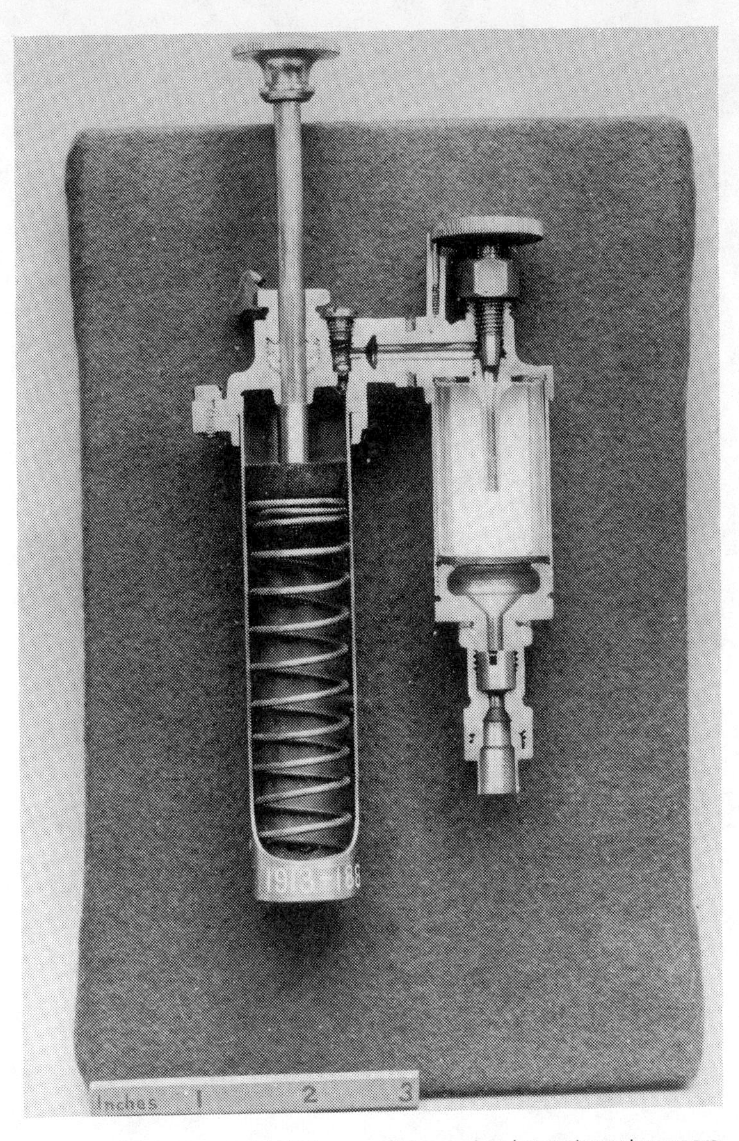

Fig. 17-3. A plunger-type lubricator was often used on buggy-type steam cars. This one clearly shows the plunger with its inverted cup-type leather washer, together with ball and spring outlet valve and the taper seat regulating valve. The oil flow can be seen in the glass sight feed which has a brass protector and sealing washers top and bottom.

Remember when setting up a newly-restored plant or lubricator, *it is better to have too much oil feed than too little.* It is always easier to slightly reduce the feed than to have to deal with the results of insufficient lubrication, which frequently results in wrecked bearings or worse. The cure for this is a return to square one, which is not good for one's equanimity.

In the case of steam buggies, there are usually various other parts requiring lubrication. In some vehicles this was simply a matter of going around with an oil can, or filling a number of Stauffer cap (screw-down) oiling units with grease. The earlier cars often either had completely open mechanisms or a simple non-oil-tight dust cover. Only later did oil-tight casing come into vogue. In this case, lubrication of the moving parts and driving mechanism was by means of a quantity of oil poured into the cover and into which the gearing of the drive dipped, scattering it around in all directions inside the cover.

In the case of engines with roller or ball bearing big-ends and main bearings, even these bearings often relied solely on an oil can, or later splash lubrication, rather than the more complicated and elaborate devices previously described.

Lubricating with an oil can is a chore necessary before starting up each day, and after each couple of hours or so of operation. Do not forget other moving parts away from the actual engine—such things as the drive to the pumps, steering mechanism, brake linkage, the pedal pivots and any hand pump linkage. All this tends to make a messy unit, but increases its life immeasurably.

Chapter 18 Engines: Component by Component

"Steam engine" is a generic term which is applied, often wrongly, to a great variety of machines ranging from a complete railway locomotive to the little device powering a child's toy boat.

Outside the scope of this book are the steam-powered rotary motors known as *turbines*. These are currently found in ocean-going ships and nuclear power generating stations. Should you be contemplating the restoration of one of these (and how you explain the presence of a 10,000 hp machine in your backyard to the neighborhood is your problem), quite frankly, you are on your own. Ignore the little man in the white coat peering at you from the background in eager anticipation of another patient!

The engines under discussion here are those within the capability of the average practical enthusiast, or small group, not needing the services of a calculator to keep track of the number of zeros on temperature, pressure and speed ranges. Our reasonably sized beast, when reduced to simple terms, produces rotary motion from steam under pressure by means of a reciprocating piston in a cylinder driving a cranked shaft through an arrangement of rods.

An integral part of the engine is an arrangement known as "valve gear" which operates a valve or valves causing the steam to be admitted to and exhausted from the appropriate parts of the cylinder at the correct time in the cycle. The valve gear incorporates easy adjustment to the timing of the valve sequence, as this is one way of altering the power output and, if the timing is altered far enough , the direction of rotation of the engine will be reversed.

Simple, Compound and Double-Acting

The numbers of cylinders and their dispositions are many and varied, but the most usual type to be encountered is a one- or two-cylinder, double-acting simple engine with the cylinder or cylinders mounted either horizontally or vertically. (See Fig. 18-1.)

Some explanation of the terms used may be helpful. "Simple," applied to a steam engine, is not an expression of its complexity but an indication that the steam is used in one simple power cycle, then exhausted. A "compound" engine, on the other hand, after using the steam in its first cylinder exhausts to the inlet valve of an adjacent or another cylinder of the same engine, working at lower pressure. This process can be continued up to three times to extract the maximum power from a given quantity of steam.

Many factory and traction engines were two-cylinder compounds, although three-cylinder (triple-expansion) engines were used, particularly in marine work. The cylinders are referred to as high pressure, inter-

Fig. 18-1. This twin-cylinder vertical engine has Hackworths valve gear and a multi-grooved flywheel rim from which belts can transmit power to whatsoever it is desired to drive.

mediate pressure, and low pressure. At each stage, a size increase in diameter is made to compensate for the reduction in pressure.

The term "double-acting" means that, unlike the auto engine, the cylinder is closed at both ends and steam is applied alternately on both sides of the piston. For an engine to be self-starting at any position, the minimum required is two cylinders, double-acting, with a crankshaft having throws 90 degrees out of phase with each other.

Common Construction Details

To deal with the construction, it is worth pointing out that, despite the amazing variety of types of engine, the majority of components are common to all. The main differences are in the type of steam distribution valve or valves employed and the design of the valve-operating gear.

Take as a common example a horizontal twin, high-pressure, double-acting engine of the type found in a typical steam car such as a Stanley. The cylinders are an iron casting incorporating a chamber between the two cylinders containing two distribution valves, one for each cylinder. This chamber is usually referred to as the valve chest and has a cover plate screwed onto it (in other designs, bolted) which must be steam-tight.

The valves are what are known as D-type, a reference to their approximate shape. They are slide valves and cover three ports or slots in each cylinder wall, the outer two of which communicate with the ends of the cylinders by means of suitably-shaped passages.

The center port, which is longer than the other two, is the exhaust and leads to the exhaust pipe. On moving the valve one way or the other, one of the end, or steam, ports is uncovered, allowing the steam from the valve chest to enter the cylinder and to apply pressure to the piston. The movement of the valve at the same time connects the port at the opposite end of the cylinder to the exhaust outlet by means of the cavity formed in the valve itself.

It would be very wasteful for steam to be admitted to the cylinder for the full stroke of the piston, so the valve is moved to "cut-off" the admission of steam at some point before the full piston movement down the cylinder. The steam now trapped in the cylinder continues to exert effort on the piston by expansion, at the same time undergoing a fall in pressure in accordance with Boyles Law.

Valve Gear—Stephensen, Walshaert and Joy

The valve is moved by an eccentric mounted on the crankshaft at such a disposition in angularity to the associated crankpin to move the valve at the correct time. For reverse motion, a second eccentric is provided, usually adjacent to the first. The rods of the "forward" and "backward" eccentrics are connected to either end of a slotted link which is suspended on arms and is capable of being moved up and down in relation to the valve spindle to bring the drive from either eccentric to bear on the valve by means of a "die-block" located in the slot of the link. If the link is in an intermediate position, the movement of the valve will be subject to the combined motion of both eccentrics. The characteristics of such motion and movement are strictly the province of the designer and need not concern the amateur. Neither should he attempt to alter same.

The type of valve gear described above is that designed by and known as "Stephensen." It is of the type of valve gear known as "link motion." There are two other valve-gear types commonly found on engines working valves of the type previously described. These are the Walshaert and the Joy, named after their respective designers. There were many variations and complete valve gears patented during the developing years of steam engines, each claiming advantages as to economy of steam consumption, simplicity of construction, or suitability for a particular location.

The Walshaert gear has only one eccentric driving a rocking link located on a fixed pivot. The valve rod is movable up and down this link and, if above the pivot, assuming the eccentric rod is connected at the bottom, the motion of the valve will be reversed. The auxiliary movement when at some intermediate position is derived from a link attached to the piston rod.

The Joy valve gear (See Fig. 18-2.) dispenses with eccentrics altogether, obtaining motion from the engine connecting rod. The method of transferring the motion to the valve and the means of reversing are somewhat different. There is a slotted link with a die-block, or a bar enclosed by a buckle which is mounted on a pivot and capable of being partially rotated. The motion from the connecting rod moves the die-block or buckle back and forth along the link. Alteration of the cut-off and reversing depends on the angularity of the link in relation to the valve spindle and driving link, the motion being transferred through approximately 90 degrees at the link.

The Joy gear is illustrated driving the other commonly encountered distribution valve—that is, the piston valve which, as the name suggests, is like a piston working in a bored cylinder. In fact, the valve is like two pistons mounted a short, fixed distance apart on the piston rod so that they cover the steam ports.

Piston Valves

The action of piston valves is similar to that described for the slide valve, but as the valve heads seal off the bore, provision is made to supply steam to the valve chest at both ends. This arrangement is known as "outside admission." A slight re-arrangement of the exhaust port, moving it from the center to the outside ends of the valve chest, enables live steam to be admitted only between the valve heads. In order to accommodate this, the timing of the valve has to be reversed from that required for the outside-admission type. This "inside admission" has the

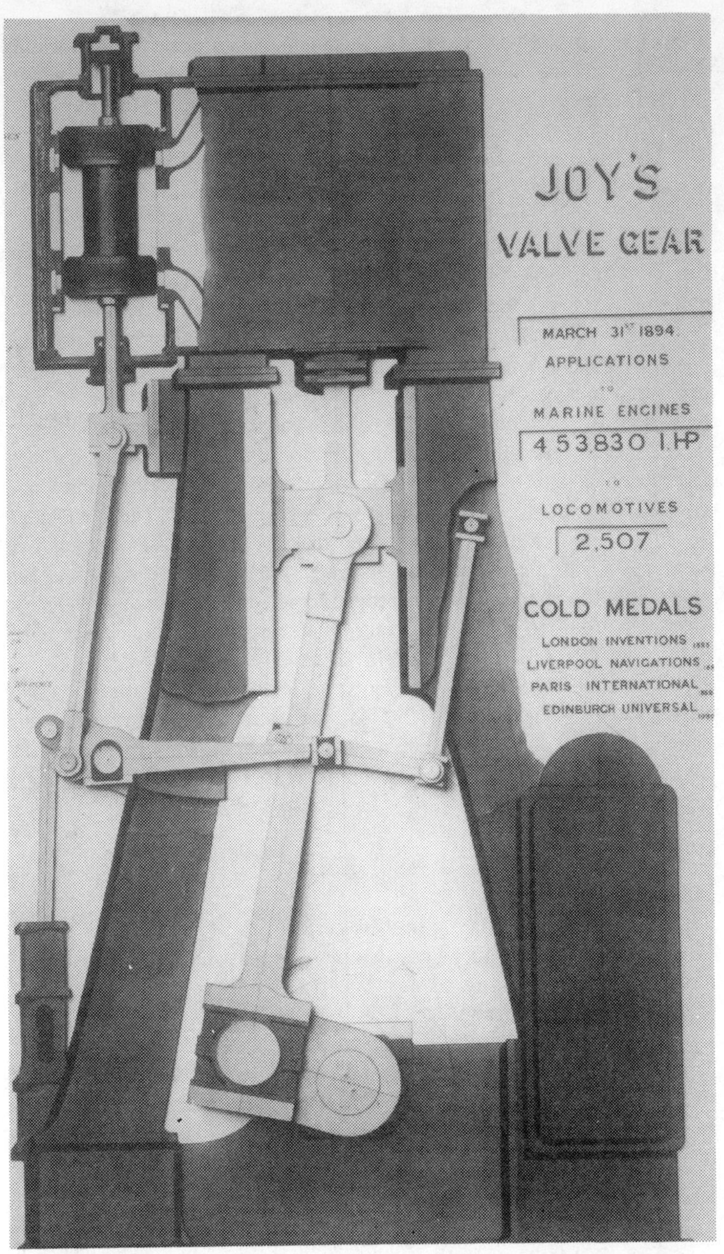

Fig. 18-2. A clear drawing of the mechanism used in the Joy valve gear, which is shown operating a piston valve of the double-headed type. Here the inlet steam enters at the center of the valve and, thus, only the exhaust pressure has to be contained by the valve rod gland and its packing.

advantage of not subjecting the valve spindle and its gland to the working pressure, only the comparatively low pressure of the exhaust steam. The effect of steam pressure on the valve was one of the reasons that the slide, or "D," valve was superseded by the more expensive piston valve.

The slide valve (See Fig. 18-3) has quite an area of flat surface in contact with the port face, and it is obvious that the whole area of the back of the valve is subject to the full steam-chest pressure which is then concentrated onto the area of contact. This close contact is necessary to ensure the steam-tightness of the valve, but absorbs power to move the valve and therefore, requires the valve gear to be of appropriate dimensions.

The piston valve, on the other hand, seals by means of rings and, because of its double-headed configuration, is subject to no unequal pressure. In fact, the force required to move a piston valve is only one-eighth of that required to move a slide valve under similar conditions.

To try to retain the ease of manufacture of the flat surface of the port face, many designs of slide valves were produced incorporating ideas and devices to try to reduce the heavy pressure loading. These included splitting the valve into two parts and making the back semi-circular in form and having fixed seals in the cover. With improved materials and machining, the piston valve was proved to be superior, although the slide valve was used right up to the end of commercial steam engine production.

One slide valve which enjoyed considerable success was the "balanced valve." (See Fig. 18-4.) This was normal in operation but had sealing strips, or a ring, between the back of the valve and the steam-chest cover, effectively reducing the area exposed to steam-chest pressure. (See Fig. 18-5.)

Poppet and Semi-Rotary

All the valves so far described perform the function of inlet and exhaust combined, but there are two more types which may be encountered:the poppet and the semi-rotary. Both of these provide separate valves for inlet and exhaust.

The poppet valve is very similar to that used on an internal combustion engine, but has a much smaller head. *The semi-rotary valve* incorporates a small cavity which, when the valve is partially turned, connects the steam space with the steam passage. The axis of this valve is at right angles to that of the main cylinder.

Both these types are operated by semi-rotary movement, directly in the case of the rotary valve and via a rocker or cam (See also Fig. 2-3.) in the case of the poppet valve. All the valve gears so far described can operate these types of valves, but usually they are operated by a simpler direct drive from the eccentrics. In the case of the exhaust valves, this is truly direct; but the drive to the inlet valves incorporates an adjustable

Fig. 18-3. A "D" or slide valve, in section, shows the sealing rings at the top and the sectioned valve shuttle at one end of its travel. In this position steam enters through the center top orifice, travels around the valve and along underneath it to emerge through the left-hand passage to one end of the cylinder and, thus, to apply pressure to the piston. Note these short steam passages which reduce dead steam space.

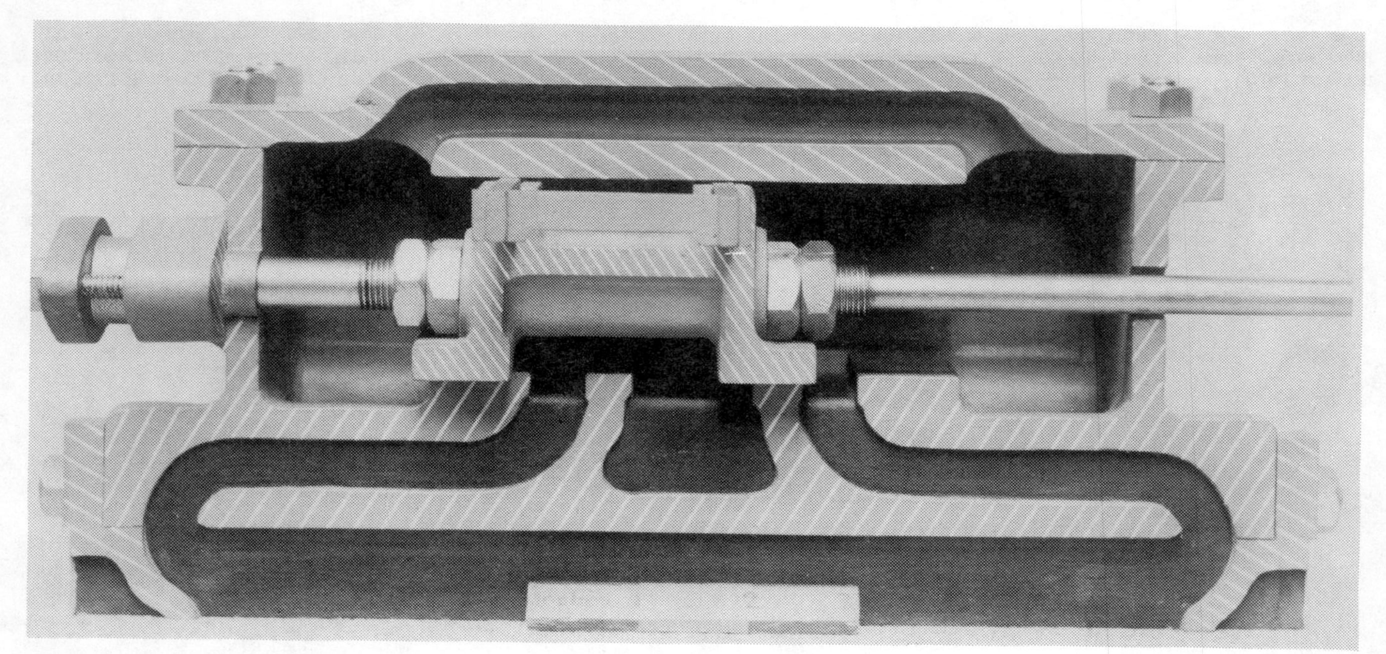

Fig. 18-4. The balanced valve is probably the most successful modification of the slide valve. Note the passage at the top of the valve chest, in the cover, to distribute steam to both ends from a single supply pipe. The screw thread and lock nuts on the spindle allow exact positioning of the valve on this spindle relative to the drive.

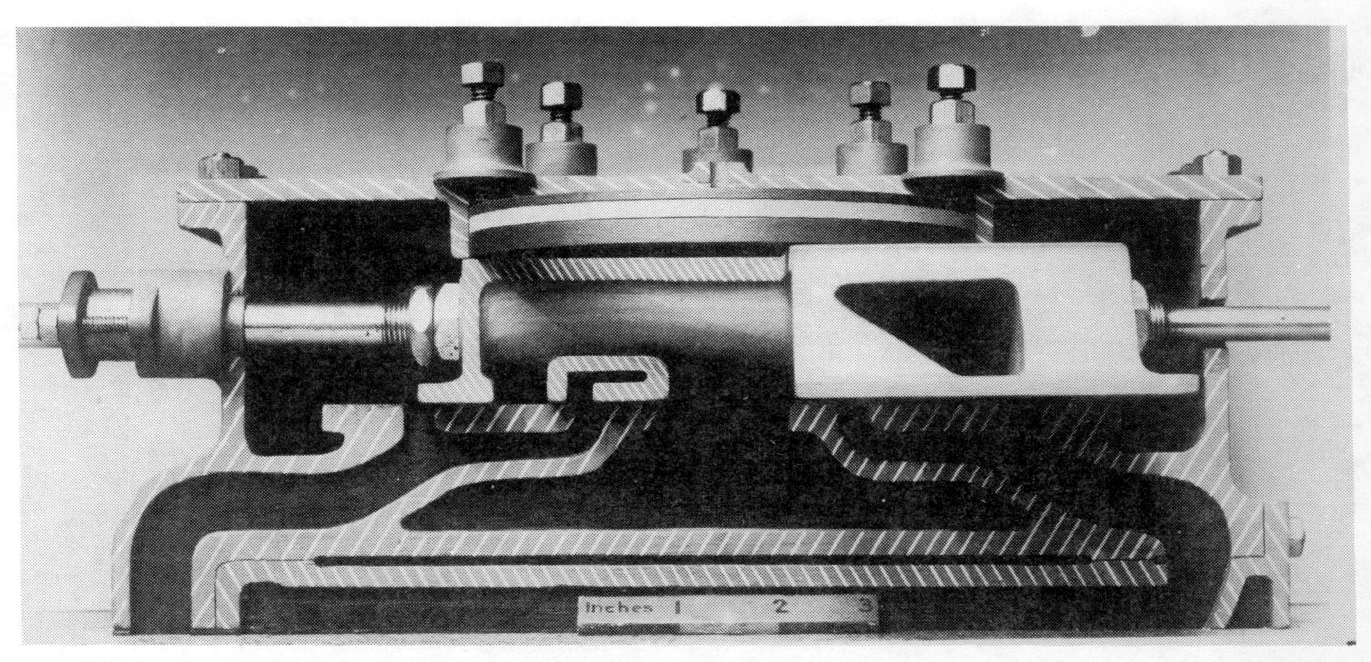

Fig. 18-5. This balanced-type slide valve has the added complication of double ports for exhaust from the cylinder. The valve is shown at full travel admitting steam to the left-hand side and exhaust from the two ports on the right. The sealing ring and its adjustment can be seen at the top.

latch device to enable the valves to cut-off before the full eccentric rod stroke.

One other method of altering the cut-off should be mentioned in connection with the slide valve. This involves an additional slide valve operating on the back of the main valve in the valve chest. (See Fig. 18-6.) In this case, the main valve has steam admission ports extending right through it, rather than just uncovering those in the port face. The main valve is driven directly from a simple eccentric and rod, which drives it the full stroke of the valve chest all the time, ensuring as free an exhaust from the cylinder as possible. The cut-off is determined by the position of the auxiliary valve regulating the steam admission through the back of the main valve. This can either be adjusted by hand, via a screw of some sort, or worked by the governor. The usual application was on stationary engines, though some firms, for example Wallis & Stevens, employed a similar expansion valve on traction engines. This was mainly intended for use while they were at work driving machinery, rather than when actually using their power to proceed along the highway.

The subject of valve gear has been dealt with at some length, as it is a sphere of great mystery to many people and is the one part of steam engine design that is quite different from anything found in other forms of prime movers. It is the most important part of a steam engine. The more easily understood items will now be examined, together with their overhaul and repair.

Cylinders and Pistons

Commencing with the cylinder and its associated piston, it may be observed that in double-acting engines the former is closed at both ends. The cylinder is normally made of good-quality cast iron, with the end cover or covers retained by a flange with studs and nuts or, in some cases, a threaded cover is screwed into the threaded end of the cylinder.

The grained nature of cast iron tends to absorb oil and is quite well protected from rusting. On removal of the covers, there may be the appearance of much rust; but this will probably prove to be only of surface type, easily removed with emery cloth and oil.

Wear in the bore is only likely to occur after many years of running, and unless really bad, it is not recommended that you go to the expense of re-boring for the limited use a "toy" is likely to receive.

The piston is formed of a forging in most cases and has split rings, usually two or more set in ring grooves around the circumference, which form the seal in the cylinder. They are often quite wide, and if you cannot get any or have them made easily it is possible to use two internal-combustion-engine type rings, side by side in the same groove. Usually replacement of these rings will recover any lost sealing, but if the bore is so badly worn or oval that it must be re-bored, this is a job for the expert machinist. Having done this, a new piston will of course be needed and,

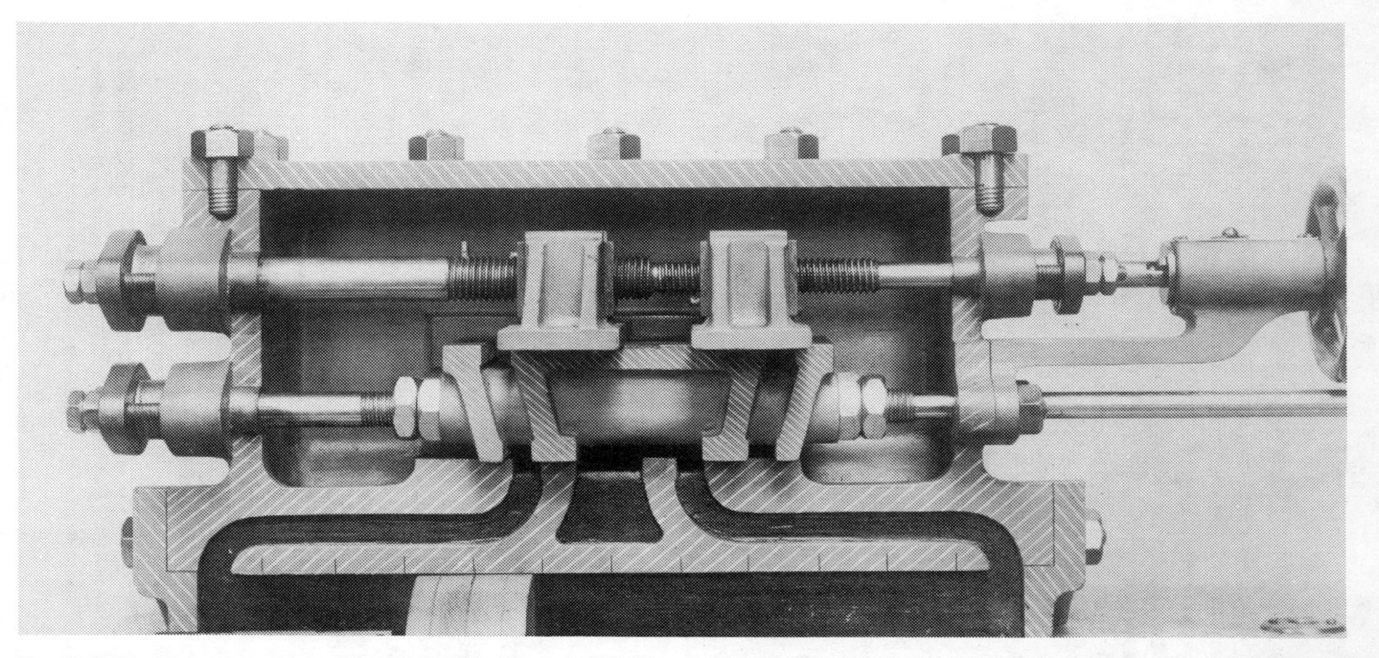

Fig. 18-6. The Meyer expansion slide valve with auxiliary cut-off valve working on the back of the main valve—this example is hand-controlled using the wheel on the right. The port on the left is open to steam admission, while that on the right is connected to the exhaust. The glands are fitted to the ends of the valve chest with setscrews and locknuts to allow for their adjustment.

almost certainly, will have to be specially made, again at your machine shop. Have a new piston rod made, as well, and both machined concentrically together at the same time. (See Fig. 18-7.)

If the original piston is to be used with new rings, it will probably be necessary to dress the piston rod to remove scoring. The piston should not be removed from the rod—usually one is screwed onto the other and then locked up with a lock nut peened over, or the end of the shaft is peened flat with the piston crown without any lock nut. Sometimes the piston rod may be tapered where it enters the piston. The reason for not removing the piston is that absolute concentricity between rod and piston is essential, and any disturbance is likely to upset this relationship.

The cylinder-end cover through which the piston rod emerges contains a bush and a stuffing box and gland to seal this rod. The bush will almost certainly require renewing, and, obviously, must be renewed if the rod is turned or replaced. This is a straightforward turning job, a brass or bronze bush being made up to suit the rod size. Always turn the rod first, as the size of this will depend upon how much material has to be removed, and the new bush can be made to any dimension to suit the rod. Give a small clearance in diameter of a few thousandths of an inch. Sometimes the bush is turned substantially larger and lined internally with white metal. Most people do not bother with this, but use plain bronze or brass.

When replacing cylinder-end covers, you may find packing or jointing between them and the cylinder. Use high-quality, high-pressure steam jointing. Make sure that the cover is fitted the correct way around. It may have some form of duct to line up with the cylinder ports. This may also apply to the screw-in type cover, in which case you must ensure that it screws up exactly the correct amount. Sometimes copper rings under the cover head are necessary to achieve this. Use graphite on the threads to make removal the next time easier.

Valve Overhaul

Proceeding from the cylinders to the valves, the face of these and the ports may be worn or scored on their mating surfaces. If this extends to more than a few thousandths of an inch, planing on a machine shop planer will be needed to restore a truly flat surface. For minor imperfections, a sheet of emery paper (about 200 grit) placed on a flat surface, such as a piece of plate glass, can be used. The procedure is to rub the surface of the valve on the emery back and forth until a true surface is obtained. Keep the emery lubricated with oil and the valve face flatly and tightly down on the sheet. It must not rock.

The port face should also be trued up if possible. If you can get at it, you can again use emery paper, wrapping it around a domestic iron and rubbing this back and forth over the port face. The iron's handle and ground base make it an ideal implement for this purpose. Some engines, for example the Stanley, have the port face buried in the valve chest and

Fig. 18-7. This horizontal engine has Corliss valve gear. Note the split brasses type, big-end bearing with adjusting shim. This engine has a large cast iron flywheel of the type that must always be carefully inspected for any flaws or cracks. The section of the cylinder shows the piston held to the piston rod by a nut and split pin. The crosshead has a single lower guide.

Fig. 18-8. This sectioned piston valve of double-headed type shows the wide one-piece piston rings. The screw threads on the valve spindle, with lock nuts, allow a measure of adjustment of the valve position relative to its spindle. The spindle bearings and gland are bolted to the valve chest covers which are, themselves, flanged and bolted to the chest. The lower ports at each end of the cylinder alternately become inlet and exhaust.

such a procedure is not possible. If such a design has badly scored port faces, the only place to deal with this is at a machine shop with an extending arm planer.

On the balance-type valve, the sealing strips of ring can be replaced using cast iron or bronze, not forgetting the flat springs which keep the seals pressed up to the cover.

Piston valves are not likely to need much renovation, as they are not subject to the same thrusts as the main pistons. Replacement of the rings is all that is likely to be required. The bore in which the piston valve slides is usually fitted with a liner, as it is easier to cut accurate ports in these before insertion. The same remarks as previously given on main piston rods apply to the valve rod or spindle.

The piston and valve rods moving in and out of the cylinder/valve-chest assembly pass through glands which require packing to try to prevent the steam escaping through them, rather than applying motion to the engine. Generally, the glands form part of the bush in the cylinder cover or the bearing of the valve spindle. (See Fig. 18-8.)

The usual system is to counterbore part of the gland/bush assembly rather larger than the piston rod or valve spindle and into this counterbore "stuff"—some reasonably soft material capable of withstanding steam and heat which can make a tight seal around the rod. To hold this material in place, a small flanged bush is then usually placed on the outer end of the rod and screwed up into the gland counterbore on top of the packing material by a cap or union nut which is screwed onto a thread formed on the outside of the outer end of the gland. Alternatively, in some designs, the cover is retained by a flange through which pass studs with nuts.

The packing material may be a series of cast iron or white metal rings, often in chevron formation, split to enable them to be placed over the rod. Following experiments in recent years, a substance called P.T.F.E. is being increasingly used for such packing duties.

P.T.F.E. is an inert plastic material with good heat-resisting properties and a very low coefficient of friction. This can be obtained in square form to the same sizes as conventional packing and requires only a sharp knife to cut it into suitable rings. A shredded variety will prove useful for very small glands, not only in engines, but also for small cocks and shut-off valves.

Sufficient packing should be inserted into the stuffing box to almost fill it and then some compression applied by lightly screwing up the gland. Final adjustment should be carried out with the engine in steam, care being taken not to overtighten the gland. In fact, just a slight seepage or wisp of steam should be visible on the reciprocating rods.

Valve Linkage Overhaul

Having dealt with the bits and pieces within the cylinder block area, the valve-operating linkage should then be checked over. With the exception of the expansion link, all the moving parts of this are pivoted on plain pins in bushed holes.

If any appreciable slack is present in any pivot, new pins and bushes can easily be turned up on a lathe and fitted. The slot in the expansion link may be found to have worn wide at the point where the die-block works when the engine is operating normally. It may be possible to carefully enlarge the rest of the slot to the widest dimension of the wear and make a new die-block a little wider to be a good fit. However, if the link has to be renewed, it should be cut from either an oil hardening steel or mild steel and subsequently case-hardened. Be very careful that the radius of the slot is made exactly the same as the original, or the valve gear may jam in operation.

The eccentrics are those odd, off-center devices that operate the valve gear by imparting a forward and backward motion. They can also be used to operate pumps.

In one design, the eccentric is enclosed by a strap which is split in half, these halves being bolted together. The strap is located by means of a groove in which a tongue, or raised part of the circumference of the eccentric, fits. Slackness of the strap can be reduced by removing shims from between the bolted holes. If no shims are present, carefully file down the mating surfaces of the strap-bolt areas. Any such filing must be absolutely flat. This may sound elementary, but try experimenting on a scrap piece of metal to see if you can file flat. It is about ten to one against that you cannot. It is one of the most difficult arts to learn.

Fig. 18-9. This Stanley steamer direct-coupled engine shows spur gears on the rear axle and engine crankshaft, with threaded rods for mesh adjustment. On the left is the steam main after the superheater, lagged; just visible is the oil feed pipe for cylinder oil, leading into it. To the right of this pipe is the ball-bearing-type crosshead of this engine.

Another design uses ball bearings between the eccentric and the eye of the operating rod. In this case, the balls probably need replacing if the engine has been lying unoiled for any length of time. Hopefully, the tracks or races will be serviceable. If not, you will have to try to find a machine shop that can tackle the job. These races will be hardened, thus they will need grinding true and new oversize balls fitted.

Crossheads and Slidebars

Another item which may need attention is the crosshead. This is a block of some sort that acts as a pivot between the piston rod, at its outer end, and the small or little end of the connecting rod.

The block may be of various shapes, but in all cases it operates in a guide, or guides, which withstands the angular thrusts from the con rod and transmits the motion into a parallel movement relative to the cylinder.

By inserting a pinch bar under the crosshead and levering, any appreciable movement will show up. If there is too much movement, it may be possible to close the guide bars up to the crosshead by removal of shims or packing pieces. If, however, wear is too bad for this, the slipper surfaces of the crosshead must be refaced with bronze or white metal linings.

Some buggy engines, for example the Stanley, employ rolling balls between guides and crosshead. (See Fig. 18-9.) These balls should be renewed if at all worn or pitted: usually there is some form of adjustment to close up the guide rails. If the grooves in which the balls run, either in the guides or the crosshead, are badly damaged, the only answer is to have new ones made up by a machine shop. This is normally beyond the scope of the amateur. Do not attempt to grind out any imperfections yourself as the surfaces are case-hardened and you are quite likely to break through the case-hardening.

The pivot pin is known variously as the "gudgeon," or "wrist" pin or just the "little-end." The pin is plain, but the bearing in the small end of the con rod may be a bush or a split brass bearing with a means for adjustment (either screw or wedge) to take up wear.

Engine Bearings

Crankshaft bearings—that is, main and big-end bearing—may be ball- or plain-metal types. Usually the ball bearing types were found in the smaller sizes, particular in road buggy engines. If there is the slightest sign of wear in these, the easiest thing to do, if possible, is to replace the whole race. Sometimes you cannot do this—for example, in the case of the big-end bearing where the outer race is part of the con rod. Then replace the balls and, if possible, the inner race, and hope the outer is not too bad. A steam engine does not have the shock loading of an internal combustion engine and, thus, does not need quite so perfect bearings. If the outer race in this type is too bad to use, your only hope is to try to

Fig. 18-10. Detail of the cylinder, valves and governor of Boulton and Watt's rotative beam ("Lap") engine of 1788—a very early example of the Watt governor on a beam engine. The governor collar is above the ball weights and, from this, the linkage goes to control steam admission.

persuade a first-class machine shop to reclaim the same for you, but it is a very difficult job.

The more common type of bearing found on steam engines is the plain-metal type, either direct bronze or bronze shells white-metal lined. These are, almost without exception, in two halves and usually have shims that can be removed between the two halves to take up wear. If no such shims are present, the mating surfaces can be filed slightly. Again, take care to set these flat.

If it is necessary to re-metal, this can be done by anyone sufficiently skilled to have reached this point in the restoration work.

Proceed as follows: Remove all the old white metal from the brasses and clean the surfaces thoroughly. Warm the brass and tin using ordinary soft-solder and flux, such as Bakers Fluid. Then clamp the halves together as they would be in the engine (the halves should be numbered).

Measure the shaft or crankpin and turn a piece of wood (the harder the better) to a slightly smaller diameter and long enough to go through the width of the bearing. Mount this vertically on a stout piece of flat hardwood, and lay the clamped brasses to that the wooden core is exactly central.

The annular gap formed is then filled with molten white metal after again warming the brasses thoroughly. After cooling, the bearing should

Fig. 18-11. The Lindley governor of 1884 is shown mounted directly on the steam valve which it operates by the descent of the central spindle as the half-round weights flyout under centrifugal force. The screw on the left adjusts the speed at which the governor operates the steam valve.

be set up in the lathe and bored to the size of the shaft. The main bearings should be mounted in the engine frame and line bored.

After boring, the shaft should be inserted in the bearings with engineer's blue smeared on the bearing surfaces. Rotating it will then mark the white metal on its high spots. Remove the bearings and hand scrape these high sots very carefully with a sharp scraper, then try again. This process will probably have to be repeated many times before a perfect bearing surface is achieved.

To re-metal a flat surface such as a crosshead, a wall of fire clay should be built up round the edge of the required area, thus forming a trough into which the molten white metal can be poured.

Before attempting to re-metal a crosshead, the piston rod must first be removed. This may be screwed in and locked with a pin in a hole drilled right through crosshead and rod,or it may be just a taper plain hole and taper on the rod, retained with a more substantial cotter. Upon rebuilding, this pin must be hammered securely home. In fact, check it even if you do not dismantle.

Governors

Most stationary engines used for driving machinery will be fitted with some form of governor. This is usually identifiable as a vertical shaft driven from the engine and fitted with two or three levers pivoted from the top. On these levers are mounted steel balls with connections to a collar underneath. (See Fig. 18-10.)

The duty of this device is to maintain the engine running at a constant speed irrespective of load. In operation, the vertical shaft spins as the engine runs and, in so doing, centrifugal force persuades the balls to fly out and, in the process, move the collar up the shaft from its "at rest" position. (See Fig. 18-11.)

This collar linkage is provided to operate a steam valve or the valve gear cut-off point. Any increase in the desired speed causes the balls to fly out still further, raising the collar further and shutting off the steam supply. As the speed falls, the converse happens, and the steam valve is caused to open again, restoring normal operating speed.

This may sound cumbersome, but in operation speed control is fairly rapid after a change in circumstance. Maintenance consists of dealing with wear in the various linkage pivot points by making new pins and bushes as required. Nearly all of these components are quite small and well within the capabilities of the amateur to reproduce on a lathe.

Most stationary engines will be fitted with a flywheel, some of these being very large and heavy. Examine these carefully for any flaws or cracks; if such exist, the wheel must not be used in this state. A wheel of this size and the large peripheral speed involved can do a great deal of damage if it breaks up when running at speed. The wheel must also run absolutely true without any wobble. Make sure any keys or retaining bolts are tight, replacing any that are the slighest suspect.

Chapter 19
The Other Car Parts

Where the restored steam engine forms part of a motor car, it is helpful to have other associated mechanical devices such as wheels, steering, brakes, etc. (See Fig. 19-1.)

The restoration of such items is covered in other TAB books such as *Automobile Restoration Guide* (TAB book No. 2001) *Car Interior Restoration* (TAB book No. 2002), and similar books. Treatment in this book will be of a superficial nature, researching as necessary, rather than going into any great detail, serving more to remind the reader of what he has to do.

Many steam cars have a steel frame or chassis, but a number do not, relying instead on the main bodywork as a support for the various mechanical devices. In these cases it is preferable to first restore the bodywork to at least its basic form, before tackling the mechanical work. This is the reverse of normal procedure. With a conventional steel frame it is normal to complete the mechanical work first, continuing this to the point of having a running chassis that can be tested for correct functioning before starting on the coachwork.

The steel frame may be of channel or tubular construction—if the former, probably riveted together; if the latter, almost certainly brazed in motorcycle fashion. If you have severe corrosion with which to contend, it is easier to get new tubes than a new channel section, but rivets are easier to replace than brazed joints; the main problem with these is removing the old tubes. In the case of rivets, many people replace these with high tensile bolts which can be tightened at a later stage if required. If a more authentic appearance is needed, the heads and most of the nuts can be turned in a lathe to a dome shape, leaving only a very small portion of the hexagon for the spanner. Dealing with slack rivets at a later date is not at all easy.

Fig. 19-1. The interesting bits of a Locomobile steam buggy, circa 1900, in this partly sectioned view, show the internals of a cylinder very clearly, together with the crankshaft and its associated connecting rods and eccentrics. To the right of the reflex-type boiler water gauge are the two back-up test cocks. The lower should always give water when opened, and the top, steam when the fire is lighted! Horizontally below the reflex gauge is the steam automatic valve that shuts off the fuel when working steam pressure is reached.

De-Rusting

For de-rusting we have found the only satisfactory method is to do so electrolytically in a caustic soda bath, at about 2 percent solution. Immerse the rusty part and connect to the negative of a DC supply. The positive should be connected to a surplus steel plate, dropped in the bath to make an anode, and of about the same area as the part being de-rusted. The voltage is not critical, but usually is somewhere between 12 and 48. Ideally, the current should be variable to provide a uniform current density no matter what size part is being de-rusted. A large car battery charger giving, say, 20 amps at a nominal 12 volts works quite well, leaving each item immersed as long as is needed to convert the rest back into hydrogen, oxygen and iron powder. These times will be found by experience.

As a guide: With the above voltage and current, an average front-axle beam will take about three hours. No harm will occur if a longer time is given. After removing, wash thoroughly in running water, dry completely and immediately paint with a rust preventative paint. The newly-cleaned metal will rust very quickly. If, after drying, the metal is not bright, a brisk rub with a wire brush will remove any remaining iron powder.

Do not immerse any aluminum parts in a caustic solution.

It is as well to adopt a pattern of restoring mechanical items. Start with the main frame, if there is one, then the springs (although these are steel, do not electrolytically de-rust them, as it will destroy the springing qualities of the leaves), followed by the axles, the axle cases, the hubs and the steering gear. If you get stuck at any point, progress to the next item. It will often be found that one is waiting for a supply from an outside source, such as a ball race, to complete one particular part. Thus, it is worthwhile ordering any parts you may need as soon as you know their extent and, if any work has to be farmed out to outside sources, getting this in hand as soon as possible.

Wheels and Brakes

Wheels and brakes can often present problems—brakes, because in early vehicles they do not work, and wheels, because they either do not exist or are in a very bad state. It is recommended that the braking system be examined before making any decision about wheels. Many early vehicles in the steam-car field had their brakes inboard from the wheels alongside the axle gear. (See Fig. 19-2.) If these worked at all, their effectiveness was marginal. Generally, either by design or accident, they ran in oil, and an oil-running contracting brake is not the most efficient means of slowing a motor car. They do eventually bite, but frequently only after the vehicle has been stopped by contacting something rather solid.

It really is not too much of a departure from originality to add brakes of *similar design* to those already fitted further out at the wheels, on the back wheels only in the case of an early car. If these brakes are to be of any value, it is helpful to affix the new drums to the wheels or the axle. In the case of wood-spoke wheels, usually a relatively simple bolt-on operation is enough. However, in the case of wire-spoke wheels it will usually be necessary to alter the rear hubs to form a brake drum integral with the spoke flange on one side.

It is obvious that it is very much easier to fit wheel brakes to wood-spoke wheels than to the wire-spoke type, if not originally fitted. If the original specification quoted only one type, it would be foolhardy to fit something different. Except in minor details, or for some specific safety or legal requirement, one should attempt to restore as accurately as possible to original design. However, many makes offered customers a choice of specification in items such as wheels.

For example, the early Stanleys had tiller steering (up to 1904) and, by the standards of the day, were quite fast, attaining 45 mph under favorable conditions. Unfortunately, it is almost impossible to straighten a bent-wood-spoke wheel without rebuilding it and this is a job for a wheelwright, a species of human being not often encountered. (See also Figs. 2-5, 2-6 and 2-9.)

If you have tiller steering and do not have perfect wood-spoke wheels, it is worth finding out if wire was an option for your particular vehicle. New clincher-type rims are available in certain sizes. (see Appendix B.)

When rebuilding the running gear, do not forget to take into account any alterations to the boiler or burner from original design. This may be necessitated by government or insurance requirements, or simply by availability. If you have to use a boiler or burner larger than original, you may have to increase the clearances of the body from the rest of the

Fig. 19-2. This Stanley rear axle shows differential and band brakes. From the early 1900's, a similar layout was used by many marques.

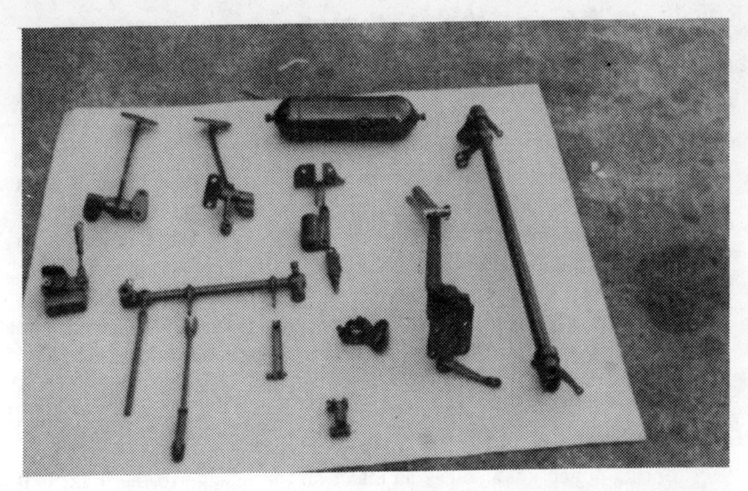

Fig. 19-3. Selection of pedals and associated control parts for a light steam buggy.

running gear, either by blocking between springs and body or, if greater weight is involved, by having the spring leaves reset and extra leaves added. These alterations may, in turn, require modifications to the engine mounting or to the transmission.

Consider also the various controls. They should be maintained in original positions as far as possible; but, again for safety reasons, it may be advisable to incorporate extra valves or safety valves and, occasionally, move a control so that it can be more easily reached. (See Fig. 19-3.)

Remember also that there are rules and regulations regarding fenders, windscreens, lights and brakes which usually did not apply when the vehicles were made, but which may now be obligatory.

Frequently, the item giving most trouble is the burner. Make sure this is fixed in such a way that it can be removed easily. Then, hopefully, you will not have to do so.

Final Drive

Normally, one does not encounter clutches or gearboxes in steam cars. Exceptions are where provision has been made to disconnect the engine from the drive to the wheels to enable the engine to run free to operate pumps for fuel and water feeds and, in the case of later vehicles, to turn a generator for electric current. In most cases, the engine is coupled permanently to the rear axle, either directly by spur gear, or via chain and sprockets or a propeller shaft. (See also Fig. 2-10.)

In the case of spur gears, the mesh should be adjusted so that the teeth on each gear make contact in the mid-portion of the tooth- profile. Do not mesh so tightly that the tip of one tooth beds right down into the root of the teeth on the adjacent gear.

In the case of chain drives, check the sprockets for wear, particularly hooking, and the chain for slackness in the links; replace if in any doubt. Even if the sprockets are not too badly worn, they soon will be if you re-fit a chain which is worn and allows its rollers to ride up on the sprocket teeth. Contrary to the remarks on spur gears, for chain drive the roots of the cog wheels should be filled by the chain. Chain tension depends upon design and length, but an average figure for free play would be the ability to push the chain into a half-inch depression from the dead taut length, to which should be added any extra slack required if deflection of the cars springing tightens the chain. This, of course, depends upon the design of the car. With a properly worked out suspension and drive layout, the tension in the chain will not vary as the springs are compressed. Few manufacturers reach this ideal state,however.

Brake Systems

Spring deflection also has an effect on the linkage to the brakes. If you have modified the brake system, you will have not only the original linkage to overhaul, but also need to make new linkage for the additional or modified brakes.

Many early vehicles failed to ensure that their brake linkage was designed and fitted in such a way that movement of the road wheels relative to the body did not result in movement of said linkage. The unfortunate effect was that when applying a brake, either by hand or foot, one could feel considerable oscillation on the lever resulting from poor geometric layout. Try to design your new linkage and modify the original connections so that the arc traversed by the brake rod is approximately the same as that through which the axle moves relative to the body. In other words, the brake rods should have a pivot point approximately in line with the pivot of the springs.

Mechanical advantage is also important, particularly where brakes are not very good and considerable pressure is needed to effect retardation. If the amount of travel needed to apply the brake at the drum can be reduced to the minimum, one is able to considerably increase the mechanical advantage of the pedal or hand lever without running out of movement of the latter.

It may also be necessary to devise a form of parking brake for cars not so fitted. Usually, the easiest and least alteration from the original in such cases is to arrange a device for holding the existing lever or pedal in the "on" position. Check with your local regulations that such a method is permissible. Some misguided authorities require even antique cars to have two entirely independent systems of braking.

Such problems only exist, of course, in early cars. Later vehicles always have two sets of brakes, often hydraulic.

Chapter 20
Steam Engine Operation

Before attempting to operate any steam plant in public, make sure you are insured. If you kill yourself, in all probability most people will be delighted. However, if you kill or hurt a third party, you are not likely to be the most popular person in town and, if not insured, may well find yourself bankrupt.

Mishaps with steam can happen only too easily and sometimes do, unfortunately, with serious consequences.

The cost of insurance varies greatly depending upon the equipment, its design, and the use to which it is put. We can only suggest you talk to your friendly insurance man, or contact one of the insurance firms listed in Appendix B.

If a boiler is to make steam, it will perform this task for better if it has some water. Any old water will not do! If your local water supply is "soft," that is, it lathers easily with soap, you are lucky. Go ahead and use it. If, as is quite likely, it is what is known as "hard," thus containing bicarbonates of calcium and magnesium, it must be treated with one of the various proprietary chemicals available (see Appendix B). If you have a domestic water softener, this water can be used. Hard water will fur up the boiler tubes and tube plates like the inside of a kettle. Not only do you lose heating ability, but local hot spots can occur due to inadequate water flow around the tube plate adjacent to the fire. (See also Chapter 11 on water feed.)

It follows that the first operation is to fill the water reservoir or tank and then to fill the boiler two-thirds. If flash type, fill the tube coil.

Next, fill the lubricators, both for bearings and cylinder-oil lubrication. Then go around all other moving parts with an oil can.

Fill the coal bunker or liquid-fuel tank or tanks. In the case of the latter, pressurize the fuel system.

If you have a coal-fired burner, you now simply rub your two bits of stick together, create fire and light up. Sit back patiently until you have enough steam to start the engine, or you go to sleep and the fire goes out or, worse still, until so much steam is made that the safety valve blows off and wakes you up!

Vaporizing the Fuel

If, however, you are one of those "lucky" people with a liquid-fuel burner, your troubles are about to begin. The crux of the problem is that liquid fuel requires to be vaporized to burn satisfactorily and, thus, needs the heat you are trying to create before you have created it, in order to produce such vaporization. Unless it is vaporized, it will not emerge from under the burner plate to burn with a beautiful blue flame.

This is the reason some people use propane or a similar gas for the burner. However, this is not to be recommended unless the burner was originally gas fed, which was not very common. The trouble is that gas has a nasty habit of disappearing from its captive state and reappearing in highly undesirable places, like underneath the equipment where a stray spark or flame ignites it, and the whole engine and those around it take to the air with unaccustomed ease!

The problem of vaporizing liquid fuel is the "chicken and egg" situation, usually resolved by the use of a gas blowtorch. This is used to preheat as much of the pipework leading to the fuel jets as can be reached. (See Fig. 20-1.) An inspection hole often exists in the burner casing into which the nozzle of the blowtorch can be directed to heat up the internal pipework or vaporizing device. Depending upon the size of the blowtorch, you may well have to spend anything from ten to 30 minutes heating up these pipes until, when you open up the pilot-light valve, a vaporized stream of gas emerges without a spurt of liquid fuel! Light this gas immediately and give the pilot light time to warm up the burner and, hopefully, the vaporizing coil for the main burner, together with adjacent pipework of the main fuel feed. You can speed up this process by using your blowtorch on this pipework, but be careful not to light the pilot jet outside the burner, which can happen in some designs.

When the burner seems reasonably warm, check the pressure to the main jets; increase, if necessary, to two-thirds operational pressure, and try a quick open-and-shut movement on the main fuel valve. It is unlikely that the burner and pipework will be hot enough to have vaporized the main fuel supply completely, so a half-gas and half-liquid fuel jet will emerge. With luck, this will complete vaporization on its way through the burner plate, and light above the plate, adding heat to the assembly and helping to heat the next charge of fuel. After one or two instantaneous operations of the main fuel valve like this, it should be possible to open it up and leave open, thus getting the main burner into operation.

Blow-Back, Pumping and Leaks

Sometimes—often!—you will get it wrong and the burner will blow-back. Try to avoid standing in its way. Most times it does no more than singe your eyebrows and hair. The author spends a fortune in false eyebrows!), but very nasty burns can occur. This event also happens very frequently when the pilot light blows out and you try to re-light it. The only certain way to avoid this is to wait some time before attempting to re-light.

Depending upon the method of fuel feed, it will probably be necessary to maintain fuel pressure by hand pumping more or less continuously once the main valve has been opened until such time as the steam pressure reaches that level at which the automatic fuel cut-off has shut. (See Fig. 20-2.)

Now is the time to look around for leaks, whether of steam, water, or in the case of a liquid, the fuel. You may well find that the various pumps leak at the glands, which usually can be tightened simply. Take up no more than is necessary to almost stop the leak. Do not over-tighten, as they will take up a little with use. In the case of road vehicles, we usually incorporate a spillage tray leading into a catch tank under the gland of each pump. This saves any fuel leakage being blown around and perhaps catching fire.

Next, check the boiler water level. If this is not correct according to the gauge glass,usually about two-thirds up the glass, hand pump to this level.

Fig. 20-1. The Steam buggy with rear boiler shows main burner jets, and air venturis, with the pilot jet. Exhaust pipe emanating from the cylinder cover at the front of the engine goes up and through the chimney to use the exhaust to produce a "pull" on the flue gases. Rocking lever driven by engine crosshead operates the pump cross-shaft under the forward floor.

Fuel Feed

You are now ready to use your steam power plant!

Depending upon its nature and design, it may or may not automatically feed itself fuel and water while in use. It is most likely to be self-feeding, if it is a relatively modern stationary plant or one of the later road vehicles. In the case of the stationary plant, such automation is more likely if working a reasonably constant load.

Coal-fired and nearly all early road vehicles are unlikely to have fully automatic feeds.

The art of stoking a coal burner is one that can only be learned with practice and involves keeping a nice level fire, sufficient for the steam requirements but not so large as to be wasteful and destructive. The principle is exactly the same for either stationary or road vehicles in this respect.

Conserving Steam

With a liquid-fueled road vehicle of the fire- or water-tube (non-flash) type of boiler, it is possible to build up a reserve of steam to anticipate extra demand for hills or acceleration. There are two ways to do this: First, stop feeding cool water into the boiler, which means making sure there is sufficient water therein when traveling light so that the water feed can be shut off when it is thought the going is about to get harder; and second, increase the heat input. This can often be accomplished by increasing the fuel pressure over and above the automatic feed level by use of the hand pump. Operated quickly enough, this will usually exceed the capacity of the fuel-regulating valve and allow fuel pressure to rise above the normal pre-set figure.

Once you have reached maximum working steam pressure as allowed by the automatic steam-fuel valve, which has shut, there is, of course, nothing more that can be done in this sphere. *Do not, under any circumstances, try to increase normal maximum steam working pressure by tampering with this valve for attempting to prevent the relief valve from blowing off.* (See Fig. 20-3.)

If there is a long hill ahead, do not rush it, but amble up slowly conserving steam. Downhill, keep a close watch on the boiler water level, or you may end up, when running on a closed throttle, pumping in far too much water. It tends to build up quite quickly when not being used as steam. If you over-fill, there is a distinct risk of "priming"—that is, feeding neat water, rather than steam, into the cylinders. Water is not compressible. The only result can very easily be no ends on the cylinders, which is a very expensive event to put right. (See Fig. 20-4.)

In an emergency, if the brakes fail, you can put the valve gear into reverse and open the throttle gently. Do this only as a last resort however, as this can also push the ends off the cylinders.

Another useful way of conserving heat is to try to arrange stops for refilling the water tank when there is an easy road ahead. This gives the new water time to acquire some heat from its surroundings before maximum steam output is again needed.

From time to time when running, check that the cylinder lubricator is working satisfactorily. Most systems have some form of indicator, but some—particularly those that use a sight-glass—are not always easy to see.

Back-Up and By-Pass Systems

With a gasoline-engined car, if something goes wrong, it is not usually possible to get home without repairing the fault. However, with a steam vehicle you can often get home after a fault has occurred, particularly if you do not have too far to go. For example, a fault in the mechanical water or fuel feeds can normally be overcome by using the hand pumps. You can progress a few miles without the lubricator working. A fault in the pilot burner does not affect the main jets. Blowing off steam, if you think maximum working pressure is about to be reached and the automatic fuel valve is about to shut, will get you home quite satisfactorily. A failing of the main jets means you are left with only a little heat from the pilot, but often this will make steam very slowly once the burner is hot.

On liquid-fuel type road vehicles it is often worthwhile incorporating various valves and pipework to enable various alternatives to be used in the event of a breakdown of one part.

For example, the author's Stanley has a valve to shut off the fuel pressure air balance tank if this bursts. (It has once!) Also, the automatic fuel valve that returns excess fuel to the tank can be isolated if it fails. This particular car is fitted with a separate tank for pilot fuel, and either this or the main tank can be coupled both to the pilot light or to the

Fig. 20-2. Pressure gauges on a light steam buggy with steam up are: (from left) main fuel reading about 100 p.s.i. pilot fuel at 20 p.s.i. and steam pressure 150 p.s.i.; above pilot gauge is oil supply to cylinders and sight gauge.

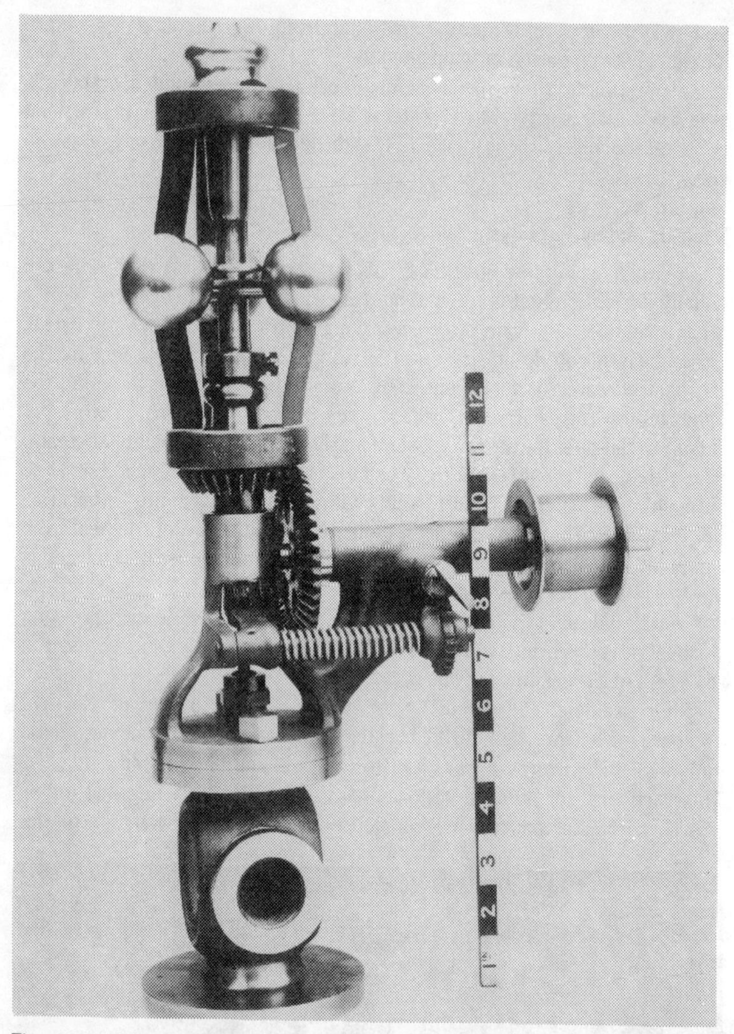

Fig. 20-3. The well-known and very widely used Pickering-type governor normally controls the main steam supply by means of the valve in the casting, shown at the bottom, which is operated by the out-swinging balls pulling down the center spindle. The drive to the governor is by means of the bevel gears and a belt from the engine, usually the flywheel, to the pulley shown on the right.

mainjets in the event of the failure of either tank. The only item one should never isolate is the steam-fuel, automatic cut-off valve. However, a device that can be by-passed, provided one is very careful when one does so, is the low-water automatic. When this is not in circuit, one has to remember that there is no safety device to cut off the fuel if the water level in the boiler gets dangerously low. A boiler running-dry can do itself immense damage.

Fig. 20-4. A close-up view of the cylinder end of a horizontal engine with Corliss valve gear—note the governor in which the ball weights swing outwards by centrifugal force as it is driven around by the belts, shown below, from the flywheel. It can be seen that this movement is transmitted by the collar below the balls, forward to the linkage on the valve chest. The direct drive to the exhaust valves can be seen at the bottom of the cylinder. The inlet valves, at the top, are driven via an adjustable cut-off operated by the governor linkage.

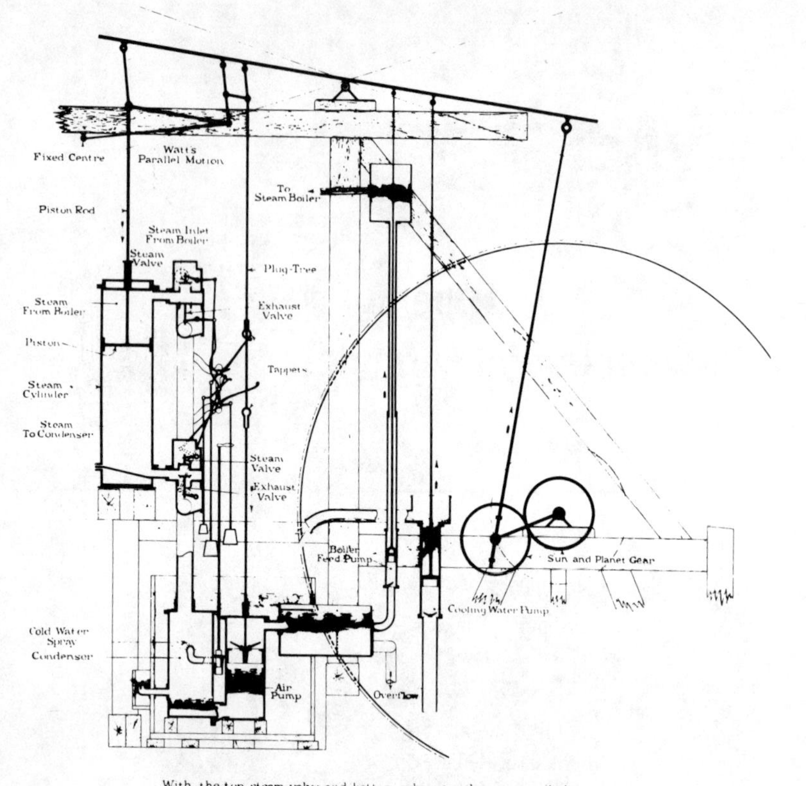

With the top steam valve and bottom exhaust valve open and the upper exhaust valve and lower steam valve closed as shown, the steam pressure above the piston and vacuum below cause it to descend. The valves are then altered by the tappets so that steam is admitted below the piston and the space above is put into communication with the condenser, where the cold water spray is working. Warm water is pumped from the condenser to the boiler. The paths of the piston-rod and plug-tree are kept vertical by the parallel motion.

Fig. 20-5. Diagrammatic view in section of Boulton & Watt's Rotative Engine, 1788.

Shutdown Check-off

When the day's work is complete, blow off all the sediment in the bottom of the boiler by opening the boiler blow-off valve at its base. Keep well clear, as high pressure steam is not good for toes.

Then close the water by-pass valve and open the water feed valve, if one is fitted. As the boiler cools, it will siphon in a fresh charge of water from the tank, ready for use next time, and save much hand pumping. Unless you remember to open the by-pass valve before the boiler has entirely cooled, it will over-fill itself. Such excess water must be drawn off before lighting up next time the plant is used, to give space for the steam.

Next, open the engine-cylinder drain cock to allow the condensate to drain away. Shut all fuel valves and release any air pressure in the fuel or balance tanks.

Appendix A
Random Useful Formulas

Engines

Theoretical Steam consumption of an engine:

$$\text{lbs. of steam per hp hour} = \frac{2{,}545}{H_1 - H_2 + T_3(\phi_2 - \phi_1)}$$

Where:

H_1 = B.T.U. per lb. of live steam

H_2 = B.T.U. per lb. of saturated steam at exhaust pressure

T_3 = absolute temperature in degrees F. of saturated steam at exhaust pressure

ϕ_1 = entropy per lb. of live steam

ϕ_2 = entropy of saturated steam at exhaust pressure

See steam tables for values of H & ϕ.

This gives an idea of the capacity of the boiler needed for a particular engine.

Usually around 34 lbs. of steam are needed to develop one horse power.

Boilers

The average boiler evaporates 3.4 lbs of water per sq. ft. of heating surface.

Rough approximation of safe working pressure:

$$\text{Safe W.P.} = \frac{(t - 2) \times S \times J}{2.75 \times D}$$

Table A-1. Specification of Fuel Suitable For Many Early Steam Cars.

	Specification	Typical
Gravity API @ 60°F		75.5
Specific gravity at 60°F		0.6836
Color	30 min.	+30
Kauri-Butanol value		29
Aniline point °F.	150 min.	155
Sulfur PPM	20 max.	2.5
Benzene vol %	0.2 max.	0.1.
Corrosion	1 B max.	1 A
During distillation	2 A max.	1 B
Distillation °F.		
LBP.	115-145	140
10%		162
50%	165-205	178
90%		204
EP	290 max.	226

The Ashland Chemical Co. of Columbus, Ohio, makes such a fuel.

For main burners, particularly on front-boilered road vehicles, completely unleaded gasoline is usually quite satisfactory and much cheaper.

where: t = thickness of shell plate in thirty-seconds of an inch

D = inside diameter of shell in inches

S = minimum tensile strength of steel of shell in tons per sq. inch

J = percentage strength of seams

J can be calculated as:

$$\text{i). } J = \frac{100(P - d)}{P}$$

$$\text{or ii). } J = \frac{2000\,(a \times n)}{s_1 \times P \times T}$$

where: a = sectional area of one rivet in sq. ins

d = diameter of rivet holes in inches

Table A-2. Heating Surface of Boiler Tubes.

External Diam. Ins.	Surface.	Ext. Diam. Ins.	Surface
5/8″	0.1636	1 5/8	0.4253
3/4	0.1963	1 3/4	0.4581
13/16	0.2127	1 7/8	0.4906
7/8	0.2291	2	0.5236
1	0.2618	2 1/2	0.6545
1 1/8	0.2945	3	9.7854
1 1/4	0.3272	3 1/2	0.9163
1 3/8	0.3599	4	1.046
1 1/2	0.3927	4 1/2	1.1781

n = number of rivets
P = pitch of rivets in inches
T = thickness of plate
s_1 = minimum tensile strength of plates in tons per sq. inch

Note two important points:

- *use the lowest value of 'J' given by these two formulas (i) or (ii);*
- *Have your calculations and the boiler checked by an expert in boilers.*

Water Tube Thickness

$$t = \frac{w.p. \times d}{55} + 7$$

Where: w.p. is working pressure in lbs/sq. in.
d = external diameter in inches
t = thickness in hundredths of an inch

For superheater tubes:

$$t = \frac{w.p. \times d}{75} + 5$$

If in doubt, err on the side of safety; increase the wall thickness.

Fuels

Weight of air required per lb. of coal for combustion

$$lbs. = \frac{\text{Calorific Value}}{10{,}000} \times 7.2$$

Calorific Values (B.T.U. per lb.)

Anthracite	15,300
Coal	15,500
Light oil	17,500
Methane	24,000

Chimneys

For stationary plants one would normally expect the chimney top area to be 1/8 to 1/10 that of the grate.

An approximate formula is:

$$A = \frac{1.5 \times G}{\sqrt{H}}$$

Where: A = area at top of the chimney in sq. ft.
G = grate area in sq. ft.
H = height of chimney in feet

Boiler Safety Devices

Area for safety valves:

$$\text{Total area of valves in sq. inches} = \frac{\text{Total heating surface}}{\text{of boiler in sq. ft.}} \times \frac{K}{p + 15}$$

Where: p = working pressure in p.s.i. (gauge)
k = 1.25 for coal fired boilers and 1.5 for oil fired boilers and boilers with forced draft

Springs for safety valves:

$$d = 3\,\frac{s \times D}{c}$$

$$D = \frac{d^3 \times c}{s}$$

$$s = \frac{c \times d^3}{D}$$

where: D = dia. from center to center of the wire (in inches-no load)
d = dia. of wire or side of square in inches (not less than 1/4-inch)
s = load on spring in lbs.

To avoid excessive compression, the spring should have a sufficient number of coils to allow, under working load, a compression of at least 1/4 the dia. of the valve.

Thus:

$$N = \frac{K \times C \times d^4}{S \times D^3}$$

$$K = \frac{S \times D^3 \times N}{C \times d^4}$$

Where: N = number of free coils in spring
K = compression in inches
d = dia. or side of square steel in 1/16-inch
C = 22 for round or 30 for square steel
S = load on spring in lbs
D = dia. from center to center of the wire (in inches—no load)

Appendix B
Contacts: Suppliers, Clubs and Insurance Companies

Boiler Makers and Repairers

AAA Retubing & Installation
40—10 Crescent St.
Long Island City, NY

A & S. Welding & Boiler Repairs
400 Botanical Square
South Bronz, NY.

Atkinson & Lawrance Inc
64 Summer,
Natick, MA 01760

Best Coil & Welding Corp
533 Remsen Ave.
Brooklyn, NY

Central Boiler Repair Co.
State Highway 35,
Oakhurst, NJ

Cruise Boiler Co.
824 N. Addison,
Elmhurst IL

Kendall Boiler Repairs
275 Third St.
Cambridge, MA 02141

Murray Frane Boilers
9901 Derby Lane,
Westchester, IL

Prime Energy Systems
5005 Jean Tacon W.,
Montreal, Quebec, Canada

Quaker Boiler Repair Co.
2012 Orthodox St.
Philadelphia, PA

Rosekilley Machinery
Box 752
San Mateo, CA 94401

Stephen Edwar
78 Franklin,
Sommerville, MA 02144

Superior Steam Generators
306 Hayward
Orange, NJ

Boiler Tubes

Boiler Tube Co. of America
McKees Rocks, PA

Chicago Tube & Iron Co
2531 W. 48th St.,
Chicago, IL

Karay (E.A. & Co.)
50 Columbus Circle,
New York, NY

Mariners-Astube Co. Inc
75 Gorge Rd.
Edgewater, NJ

Markland Scowcroft Ltd.
Bromley Cross
Nr. Bolton, Lancs, UK

Clubs

Antique Auto Club of America
501 Governor Rd.
Hershey, PA., 17033

Classic Car Club of America
P.O. Box 443,
Madison, NJ 07940

Historical Auto Society of Canada
264 Main St. Site 1, Box 6, RR2,
Rockwood, Ontario, Canada

Horseless Carriage Club of America
9031 E. Florence Ave.,
Downey, CA 90240

Steam Auto Club
1937 East 71st. St.
Chicago, IL 60649

Steam Automobile Club of America
333 N. Michigan Ave.,
Chicago, IL 60601

Veteran Car Club of G.B.
Jessamine House,
Ashwell, Herts, UK

Vintage Sports Car Club
121 Russell Rd.
Newbury, Berks, UK

Veteran Motor Club of America
105 Elm St.,
Andover, MA 01810

Vintage Car Club of Canada
P.O. Box 3070,
Vancouver, BC., Canada

Car Dealers

Antique Auto Sales
1414 Oakland Park Ave.,
Columbus, OH 43224

Antique & Classic Cars
4117 Reserve,
Missoula, MT 59801

Autoland Inc
U.S. 1 Box 126,
Hobe Sound, FL 33455

Automotive Clasics
209 Colorado Ave.,
Santa Monica CA 90401

Bills Antique Cars
Rt. 1. Box 364A,
Muskogee, OK 74401

Classic Car Investments
1701 Spring St.
Smyrna, GA 30080

Fagan Classic Cars
222 West 7th,
Topeka, KN 66603

Gephart Classic Cars
340 N. Main St.,
Englewood, OH 45322

Roaring Twenties
Rt. 1 Box 198,
Hood, VA 22723

Stanley Sales & Service
(Carl Amsley)
Rt. 2, Box 69,
St. Thomas, PA 17252

Veteran Car Sales
2030 South Cherokee St.,
Denver, CO 80223

Gauges

British Steam Specialities
Fleet St. Lee Circle,
Leicester, UK

J. Lynch
33 Loomis St.,
Little Falls, NY 13365

Nisonger Corp
35 Bartels Pl.
New Rochelle, NY 10801

Paul Sullivan
4311 Sunset Blvd.,
Los Angeles, CA 90029

Gus. Scheuer (dials only)
31 Dunwoodie Place,
Greenwich, CT 06830

Temperature Gauge Guy
45 Prospect St.,
Essex Jct., VT 05452

Vintage Instrument Restoration
(John Marks)
4 Whybourne Crest,
Tunbridge Wells, Kent, UK

General

	Specialty
Amsley (Carl) Stanley Expert: Rd 2. Box 69, St. Thomas, PA 17252	STANLEY CARS
Babbitt Pot: 216 Maple St. Glens Falls, NY 12801	RE-METALLING
Bahco Tools: Beaumont Rd. Banbury, Oxon, UK	GAS BLOWLAMP
Bassett Lowke: 59 Cadogan St. London S.W. 3, UK	SMALL STEAM FITTINGS
Butler Radiators: Box 157, Rt. 1, Butler, MO 64730	CONDENSORS
Cal's Engine Shop: P.O. Box 434, 2146 Main St. Exalon, CA 95320	RE-METALLING

Harkin Machine Shop: RE-METALLING
115 1st. Ave. N.W.,
Watertown, S.D. 57201

Heil Auto Machine Co: CONN. RODS
P.O. Box 507 4 (Blvd),
Glen Falls, NY 12801

Herbert Terry: SPRINGS
Redditch, Worcs,UK

I.M.I. Range: COPPER TANKS
P.O. Box 1,
Stalybridge, Cheshire, UK

Johns Quality Pistons: PISTONS
2662 Lacy St.,
Los Angeles, CA 90031

Mann Manufacturing Co: STEAM AUTO-PARTS
RR3. Box 196.AA
Augusta, KN 67010

Pilkington Ltd: GLASS FOR SITE GAUGES
Prescot Rd.,
St. Helens, Lancs, UK

Pollock Auto Showcase: CONDENSORS
705 Franklin St.,
Pottstown, PA 19464

Repro-Tiques: GAS & WATER TANKS
Box 130, Rt. 4,
Hot Springs, AR 71901

Speedmid Ltd: MUDGUARDS
Tame Rd.,
Witton, Birmingham 6, UK

Spinform: COPPER TANKS
Clough Rd.,
Blackley, Manchester, UK

Turner & Newall: ASBESTOS
P.O. Box 40,
Rochdale, Lancs, UK

Glass

Classic Auto Glass
P.O. Box 1. 13014 West 7th Place,
Golden, CO 80401

Henry's Auto Glass
2315 Eagle Ave.
Alameda, CA 94501

Independent Glass Co.
62 Harvard St.,
Brookline, MA 02146

Pilkington Ltd
Prescot Rd.,
St. Helens, Lancs, UK

Insurance

Condon & Skelly
P.O. Drawer A,
Willingboro, NJ 08046

James A. Grundy
500 Office Center Drive,
Fort Washington, PA 19034

Classic Ins. Agency
639 Lindbergh Way N.E.
Atlanta, GA 30324

J.C. Taylor
8701 West Chester Pike,
Upper Darbey PA 19082

Pipes, Fittings and Valves

Academy Metal Products
1 Edward J. Hart Drive,
Jersey City, NJ

All Stainless Inc
75 Research Rd.
Hingham, MA 02043

British Steam Specialities
Fleet St.,
Lee Circle, Leicester, UK

Chicago Tube & Iron Co
2531 W. 48 th St.
Chicago, IL

Coupco Mercantile Corp
570 Coronation Dr.
Toronto, Ontario, Canada

Envit Ltd. Abercanaid
Merthyr Tydfil, Sth.
Walves, UK

Hattersley-Newman-Hender
Burscough Rd.
Ormskirk, Lancs, UK

Independent Pipe Corp
Whiteman Rd.,
Canton, MA 02021

Markland Scowcroft (Pipes)
Bromley Cross Nr. Bolton,
Lancs., UK

Mindeco Corp
3345 Royal Ave.,
Oceanside, New York, NY

Neil Supply Co
700 Schuyer Ave.
Lyndhurst, NJ

Pegler & Louden Ltd
54 Brown St.,
Glasgow, Scotland, UK

Rotherman & Sons
Ace Works
Parkside, Coventry, UK

Superior Pipe Specialities
2917 S. Cicero,
Cicero, IH

Superior Pipe Specialities
2917 S. Cicero,
Cicero, IH

Shanks & Co. Tubal Works
Barrhead, Glosgow,
Scotland, UK

Slingsby (G.&.A.E.)
Cleveland St.,
Hull, UK

Waterman Machine Mfg. Co
1215 Germanstown Ave.,
Philadelphia, PA

West Jersey Manufacturing Co.
Main & Chestnut St.
Williamstown, NJ

Young (T & Co.) Centric Works
121 Barr St.
Birmingham, UK

Electro-Plating

Ace Chrome Plating Co.
728 W. Highland,
Milwaukee, WI 53233

Bill's Metal Polishing.
Davis Rd. & Camden Ave.
Magnolia, NJ 08049

Classic Custom Plating
North Wester at 71st.,
Oklahoma City, OK 73116

Davison Plating Co
845 N. State Rd.
Davison, MI 48423

Graves Plating Co
Industrial Park, P.O. Box 1052H,
Florence, AL 35630

High Grade Plating Co
1245 W. 2nd St.
Pomona, CA 91766

Hygrade Plating Co.
2207 41st Ave.,
Long Island City, NY

Modern Plating
242 So. 12th St.
Newark, NJ 07107

O'Donnell Plating
41A Mill St. P.O. Box 33,
Forest Park Station,
Springfield, MA 01108

R & S. Plating
1933 Forster St.,
Harrisburg, PA 17103

Watervilet Plating Co.
911 11th St.,
Watervilet, NY 12189

Pumps

Allied Pump Corp.
128 Harrison,
Hoboken, NJ

Ace Pump Co.
57 West 21st St.,
New York, NY

Bitzer & Co.
1330 Willow Ave.
Melrose Park, PA

Ford & Ivester Ass. Inc
290 2nd Waltham, MA 02154

Fluid Pump Service
6544 N. Milwaukee,
Chicago, IL

John Tucker Ltd.
2803 Botham,
Montreal, Quebec, Canada

Ketchum Pump Co.
34—2064 St.,
Woodside, NY 11377

Stuart Turner
Henley-on-Thames,
Oxon. UK

Stanley Service
Rd 2., Box 69,
St. Thomas, PA 17252

Tranco Pump Co.
1500 W. Adams,
Chicago, IL

Trench & Marine Pump Co.
1526 West 22nd St.
New York, NY

Weil Pump Co.
1530 N. Fremont,
Chicago IL

Tube Expanders

Chicago Tube & Iron Co.
2531 W. 48th St.
Chicago, IL

Dresser Industries Inc.
1174 W. Chestnut,
Union, NJ

Elliot Co.
1071 Bristol Rd.,
Mountanside, NJ

Goodway Tools Cn.
220 Stillwater Ave.
Stamford, CN

Wilson (Thorn C.) Inc.
21—1144 Ave.
Long Island City, NY

Weymouth Sales Co.
3733 N. Kimball,
Chicago, IL

Tires

Bills Antique Tires
P.O. Box 176, 7526 Kay Lynn St.,
Stanley, KN 66223

Coker Tire Co.
5100 Brainerd Rd.
Chattanooga, TN 37411

Denham Rubber Manufacturing Co.
P.O. Box 951,
Warren, OH 44482

Dom's Tire Sales Inc.
249 Lumber St.
Coatesville PA 19320

Kelsey Tire Inc.
Box 564,
Camdenton, MO 65020

Lester Tires
26881 Cannon Rd.
Bedford Heights, OH 44014

Lucas Automotive Eng
11848 W. Jefferson Blvd.
Culver City, CA 90230

Universal Tires
6650 Columbia Ave.
Lancaster, PA 17603

Vintage Tire Supplies
Jackman Mews Off North Circular Rd.
Neasden, London N.W. 10 England

Upholstery

Allender (Frank)
7323 Reseda Blvd.,
Reseda, CA 91335

Antique Fabric & Trim Co
Rt. 2. Box 870,
Cambridge, MN 55008

Bill Hirsch
396 Littleton Ave.
Newark, NJ 07103

Carters Cover Shop
Box 80: 800 East 6th St.
Beardstown, IL 62618

Hides Inc
P.O. Box 30,
Hackettstown, NJ 07840

Pouyat (Carl)
P.O. Box 292, 235 Lexington Ave.
Murray Hill Station, NY 10016

Sitts
Highway 23,
Churchtown, PA 17510

Studebaker Acres
580 Massillon Rd.,
Akron, OH 44306

Water Treatment

Aqua Labs
16 High St.,
Amesbury, MA 01913

Brirco Labs
1551—63 Brooklyn,
New York, NY

Continental Water Conditioning Co.
490 Hendricks Causeway,
Ridgefield, NJ

Crane Co
300 Park Ave.
New York, NY

Chemed Corp.
2301 Algonquin Pkwy.,
Rolling Meadows, IL

Liber Rich & Sons
2524 Arctic Ave.
Atlantic City, NJ

Wheels

Dayton Wheel Products.
2326 E. River Rd.
Dayton, OH 45439

Ernie De Bow.
16756 Bennett Rd.,
North Royalton, OH 44133

Elster Hayes, Oakrest Machine Shop
2110 Boda St.
Springfield, HO 45503

Coalman (Stan)
Box 1002 R.,
Morristown, NJ 07960

Fielder (Stephen)
Box 5885,
Ocean Park, CA 90405

Harry Johnson
2570 Pioneer Drive,
Reno, NV 89509

Owen Higgs
2532 Pringle Circle,
Ogden, UT 84403

Restoration Wheel Shop
30 East North St.
Waynesboro, PA 17268

Studebaker Acres
580 Massillon Rd.
Akron, OH 44306

Terra (Joseph)
17093 N. Tretheway,
Lodi, CA 95240

West London Repair Co.
5 Lancaster Rd.
London S.W.19., England

Wheel Repair Service.
176 Grove St.,
Paxton, MA 01612

Willies Antique Tires
5257 W. Diversey Ave.,
Chicago, IL 60639

Vintage Tire Supplies
Jackman Mews. Off North Circular Rd.
Neasden, London N.W. 10.England

Appendix C Directory of 400 Steam Cars

A.B.C.
Achille Philian
Alena
Allen
Alma
Altham
American
American Locomotor
American Steam
American Steam Buggy
American Steam Car
American Steamer
American Waltham
Anderson
Anderson Steam
Anderson Steamer
Anderson Whitney
Argonaut
Artzberger
Atlantic
Aultman
Austen
Austenius
Auto Loco

Baker
Baldwin
Ball
Banks
Bar Harbor
Barlow
Barrett
Barton
Battin
Battke
Bauer
Belger & Bowker
Belknap
Benson
Best
Billings
Binney-Burnham
Blair
Blaisdell
Blanchard
Blevney
Bliss
Bluffclimber
Bob Cat
Bobsover
Bohnet
Bollee
Bon-Car
Boss
Brecht
Bridgeport
Bristol
Brooks
Brown
Bryan
Bullard
Bundy

Cameron
Cannon
Capitol
Carhart
Carqueville-McDonald
Carter
Caswell
Catley & Ayres
Central
Century-
Chapman
Chautaugua Steamer
Chicago
Cincinnati
Clark
Clarkson
Clearmont
Clegg
Clinton

Coats
Coffin
Collins
Colonial
Columbia
Commercial
Comton
Connolly
Conrad
Copeland
Cotta
Crompton
Cross
Crouch
Cruickshank
Cullman
Cunningham
Curran
Curtis

Daley
Darling
Davenport
Dawson
Day
Dayton
De Dion
Delling
Derr
Detroit Steamer
Detroit Steam Car
De Weese
Dillon
Doble
Dodge

Eastman
Eclipse
Economic
Einig
Elberon
Electronomic
Elite
Empire
Empire State
Endurance
Essex
Everett
Excelsior

Fairfield
Fauber
Fawcett-Fowler
Federal
Fields
Fisher
Flinn
Foster
Foye
Frantz
Frazer
Freeman
French

Gaeth
Gage
Gagemobile
Gardener-Serpollet
Gearless
Gearless Steamer
Geer
Geneva
Gibbs
Gibson
Gould
Graham
Green Bay
Griffiths
Grout

Hackley
Halsey
Hammett
Hart
Hartkey
Hazelton
Henley
Henrietta
Herreshoff
Hersehmann
Hess
Hidley
H.L.B.
Hoffman
Holland
Holmes
Holyoke
Hood
Hoighton
House
Howard
Hoyt
Hudson Steam
Hudson Steamer
Hughes
Hunt
Hydro

Ideal
Ingersoll-Moore
International
Iroquois

James
Jamieson
Jaxon
Jeffries
Johnson
Johnson Steamer
Jones Steamer

Keene Steamer
Keen Steam
Keenelet
Kellogg
Kelly Steam
Kensington
Kent
Kent's Pacemaker
Keystone
Kidder
King
Knight
Kraft
Kramer

Lane
Lane & Daley
Larchmont
Lawler

Leach
Lear
Leroux
Lifu
Light
Locke
Locomobile
Long
Loomis
Lozier
Lutz
Lyons

MacDonald
Malden
Manchester
Marlboro
Marsh
Marshall
Maryland
Mason
Massachusetts
McCormick
McCurdy
McGee
McKay
McMurtz
McQuestion
Memphis
Mercury
Meteor
Meserve
Michigan Steamer
Milburn
Miller
Mills
Milne
Milwaukee
Mobile
Model
Moncrieff
Moore Spring
Moore
Morriss
Morse
Muir

National
Newcomb
New England
New Home
New York
Norton

Ofelt
Olive
Ophir
Orcutt
Ormond
Ovenden
Overholt
Overman
Owens
Oxford

Parker-Wearwell
Parsons
Pawtucket
Pearson-Cox
Peerless
Peet
Pennsylvania
Perry
Philion
Pierce
Pilgrim
Pittsburgh
Polhemus & Thomas
Porter
Prescott
Puritan

Ramapaugh
Randall
Rand & Harvey
Randolph
Reading
Reilly
Remal-Vincent
Richmond
Riley & Cowley
Ritter
Rochester
Rogers
Roper
Ross
Rushmobile
Rutherford

Safety
S.B.M.
Scott-Newcomb
Sears
Shaffer
Sharp
Shatswell
Shaver
Sheppee
Shillito
Simmonds
Simons
Simplex
Simpson
Skene
S & M
Smith
Snyder
Spencer
Spokane
Springer
Soames
Springfield
Squier
Stanmobile
Standard
Standard Steam
Standard Steamer
Stanley
Stanley-Whitney
Stanton
Star
Steamobile
Steam Vehicle
Stearnes
Steel Ball
Sterling
Sternes
Stewart-Coats
Stokesbury
Storch

Strathmore
Stringer
Strouse
Suburban
Sunset
Superheated Steam Car
Super Steamer
Suttle
Sweany

Taft
Taunton
Taylor
Terwilliger
Thompson
Thomson
Tinkham
Twy
Toledo
Tractomobile
Transit
Trask-Detroit
Trask-Steam
Trinity
Triumph
Turner-Meisse
Twombly

United States

Vanell
Vapomobile

Walker
Wallof
Waltham
Warfield
Warner
Washington
Watch City
Watt
Webb-Jay
Westfield
Western
West
White
Whitney
Whitney Steamer
Willard
Williams
Windsor
Wood
Wood-Loco
Wood Steamer
Worthly

Zachow & Besserdich

Appendix D
Museums for Steam

Some of these museums have permanent displays incorporating steam equipment. Others rotate their exhibits and may sometimes have steam plants on show. A few of these museums sell exhibits. Not all are open all the year. They are listed in alphabetical order by state.

Arkansas

Museum of Automobiles
Rt. 3,
Morrilton, AK 72110

California

Briggs Cunningham
250 Baker St.
Costa Mesa, CA 92606

Movieworld Cars of the Stars
6920 Orangethorpe Ave.
Buena Park, CA 90602

Colorado

Ray Dougherty Collection
Rt 2 Box 253A,
Longmont, CO 80501

Forney Transportation Museum
P.O. Box 176 1416 Platte St.,
Denver, CO 80202

Delaware

The Magic Age of Steam
P.O. Box 27 Rt. 82,
Yorklyn, DE 19736
(This has the best display of Stanley Steamers in the U.S.A. but is not open every year—check.

Washington D.C.

National Museum of History & Technology
Smithsonian Institute

Florida

Bellm's Cars of Yesterday
5500 North Tamiami Trail,
Sarasota, FL 33580

Early American Museum
P.O. Box 188,
Silver Springs, FL 32688

Georgia

Antique Auto Museum
Stone Mountain, GA 30088

Illinois

Museum of Science & Industry
57th. St. & Lake Shore Drive,
Chicago, Il 60637

Time Was Village Museum
1325 Burlington St.,
Mendota, IL 61342

Maine

Wells Auto Museum
P.O. Box 496 Rt. 1,
Wells, ME 04090

Massachusetts

Heritage Plantation of Sandwich
P.O. Box 566 Grove St.,
Sandwich, MA 02563

Museum of Transportation
15 Newton St.,
Brookline, MA 02146

Michigan

Henry Ford Museum
Dearborn, MI 48121

Poll Museum
353 East 6th St. (U.S. 31
Holland, MI 49423

Minnesota

Hemp Museum
P.O. Box 851,
Rochester, MN 55901

Missouri

National Museum of Transport
3015 Barrett Station Rd.
St. Louis, MO 63122

Nebraska

Harold Warp Pioneer Village
Minden, NB 68959

New York

Long Island Auto Museum
Meadow Spring,
Glen Cove, NY 11542

Upstate Auto Museum
Box 152, US. 20,
Bridgewater, NY 13313

Ohio

Crawford Auto-Aviation Museum
10825 East Blvd.
Cleveland, OH 44106

Pennsylvania

Pollock Auto Showcase
70 S. Franklin St.,
Pottstown, PA 19464

South Dakota

Pioneer Auto Museum
Box 76, Int. 90,
Murdo, SD 57559

Texas

Chapman Auto Museum
Rt. 3, Box 120A,
Rockwell, TX 75087

Virginia

Car & Carriage Caravan
P.O. Box 748,
Luray, VA 22835

Canada

Canadian Auto Museum
99 Simcoe St. So.,
Oshawa, Ontario, Canada

Craven Foundation
760 Lawrance Ave. West,
Toronto, Ontario, Canada

Reynolds Museum
Box 6780, Highway 2A,
Wetaskiwin, Alberta, Canada

British Isles

Museum of Science & Industry
Newhall St.
Birmingham, England

Manz Motor Museum
Crosby, Isle of Man

Myreton Motor Museum
Nr. Aberlady, East Lothian, Scotland

National Motor Museum
Beaulieu, Hampshire, England.

Index